Stillman Williams Robinson

Principles of Mechanism
A Treatise on the Modification of Motion by Means of the...

ISBN/EAN: 9783337002398

Printed in Europe, USA, Canada, Australia, Japan

Cover: Foto ©berggeist007 / pixelio.de

More available books at **www.hansebooks.com**

PRINCIPLES OF MECHANISM.

A TREATISE ON

THE MODIFICATION OF MOTION

BY MEANS OF THE

ELEMENTARY COMBINATIONS OF MECHANISM,

OR OF THE PARTS OF MACHINES.

*FOR USE IN COLLEGE CLASSES,
BY MECHANICAL ENGINEERS,
ETC., ETC.*

BY

STILLMAN W. ROBINSON, C.E., D.Sc.,

Mechanical Engineer and Expert for the Wire Grip Fastening Co.; Vice-President and Mechanical Engineer to the Grip Machinery Co.; till recently Professor of Mechanical Engineering in the Ohio State University; Member Am. Soc. Mechanical Engineers; Member Am. Soc. Civil Engineers; Fellow of the American Association for the Advancement of Science; Member Am. Soc. Naval Architects and Marine Engineers; and Member Soc. for Promotion of Engineering Education.

FIRST EDITION.

FIRST THOUSAND.

NEW YORK:
JOHN WILEY & SONS.
LONDON: CHAPMAN & HALL, LIMITED.
1896.

PREFACE.

This work aims to treat the whole subject of MECHANISM in such systematic and comprehensive way that by its aid any machine, however elaborate, may be analyzed into its elementary combinations, and the character of their motions determined.

In the classification, the *System* of Prof. Robert Willis has been followed in the main, as serving best the present purpose; and largely his names and terms as well.

The work contains the substance of lectures given in my classes during the past twenty-seven years, with such additions and amplifications as might come by reason of a somewhat extended study of the subject not only, but as brought out in connection with an aptitude for the thinking out of inventions involving more or less novel and varied forms and combinations.

Besides Willis, some of the authors whose works have been referred to with profit may be named: Prof. C. W. MacCord; Geo. B. Grant; J. W. M. Rankine; F. Reuleaux; J. B. Belanger; Ch. Laboulaye.

Some of the topics to which special attention is invited, as embracing either entirely new solutions of important questions, or previously unpublished discussions and extension of inquiry concerning them, are:

> Log-spiral Multilobes as derived from one spiral; also Proportional Sectors.
>
> Easements to Angular Pitch Lines.
>
> Transformed Wheels.
>
> General Solution of Non-circular Wheels, External and Internal, for the case of Given Laws of Motion.
>
> Similar and other Wheels from Auxiliary Sectors, Plane and Bevel.

Special Bevel Non-circular Wheels laid out on the Normal Sphere.

General Solution for Bevel Non-circular Wheels for Stated Laws of Motion.

General Solution for Skew-bevel Non-circular Wheels for Stated Laws of Motion.

Practical Rolling of Pitch Lines and Engagement of Teeth, in one Pair of Non-circular Wheels.

Intermittent and Alternate Motions for Moderate or for High Speeds; Circular and Non-circular.

Internal or Annular Non-circular Wheels.

Teeth for Skew-bevel Non-circular Wheels.

"Blocking," and Steepest Gear Teeth.

Interference of Involute Teeth of Annular Wheels.

Epicycloidal Engine and Accessories for Machine-made Teeth.

Full Discussion of Olivier Spiraloids; Interference, etc.

Cam Construction by Co-ordinates.

Form of Roller for Cams.

Solution for Cams with "Flat Foot" Follower.

Cam of Constant Breadth and Given Law of Motion.

Easements for Cams.

Solutions for Varied Velocity-ratio in Belt Gearing.

Non-circular Pulleys for Continuous Motion by Law.

Solution for Cone Pulleys.

Rolling Curve Equivalent for Link-work in General; Plane and Bevel.

Gabs and Pins for Link-work in General; Plane and Bevel.

Velocity-ratio in Bevel and Skew-bevel Link-work.

A General Crank Coupling connecting Shafts in Various Planes and Angles.

Practical Forms of Parts for Bevel and Skew-bevel Link-work.

Varied Step Ratchet Movement.

Face Ratchets and Clicks.

The aim of the work is not so much to present a history of **Mechanism**, as to treat upon the principles which underlie the various modes of modification of motion as due to the form and connection of parts, and thus to enable the inventor and designer of machines to at once solve any problem of motion of an elementary combination in a particular case.

The treatment has been mainly by graphics instead of by analysis, for three reasons: 1st, because the draftsman's outfit is usually at hand when mechanism problems arise; 2d, because these problems usually do not require the precision of analysis, the graphic method serving for problematic work as well as for delineation; 3d, because analysis, though possible in a few of the simpler problems, becomes difficult and often impossible with very slight variations of conditions, while by the graphic method all cases, whether of simple or complex statement, are solved with nearly equal facility. Hence analysis has been employed here only to establish principles the application of which subsequently might be made by the graphic method.

S. W. ROBINSON.

COLUMBUS, O., *Sept.* 26, 1896.

CONTENTS.

— — —

INTRODUCTION.

OBJECT OF PRINCIPLES OF MECHANISM .... Page 1
 How Studied. Machine Defined. Frame; and Trains of Mechanism. Machine Parts Classified. Elementary Combination of Mechanism. Motions, how Studied, **1.** *Synoptical Table of Elementary Combinations of Mechanism,* **2.** Velocity; Constant and Varied; Angular. Velocity-ratio. Period. Cycle, **3.** Revolution. Distinction into *Driver* and *Follower.* Directional Relation, **4.**

PART I.

TRANSMISSION OF MOTION BY ROLLING CONTACT.

CHAPTER I.

ROLLING CONTACT IN GENERAL............................ Page 5
 Line of Contact. Axes. Plane of Axes. Point of Contact. Line of Centers. True Rolling Contact. Velocity-ratio in Circular Rolling Contact. No Slip of Surfaces in Contact, **5.** Velocity-ratio of Non circular Arcs, **6.** Variable Velocity-ratio, **7.**

CIRCULAR WHEELS.. Page 7
 Axes Parallel. Friction Wheels. Contact between the Axes, **7.** Contact Outside the Axes. Axes Meeting, **8.** Parallelogram for Locating Line of Contact. Rolling Cones. Angle between Axes, Particular Case, **9.** Axes Crossing without Meeting. Character of Contact between the Surfaces. Form of the Rolling Surfaces. Location of Line of Contact in Plan, **10.** Location of Line of Contact in Elevation. Hyperboloidal Form of Surfaces. Longitudinal Slip of Surfaces. Error in Early History of Case, **11.** Graphic Construction of Skew Bevels. Proof of Construction, **12.** Not Practical for Friction Wheels. Longitudinal Slip Determined, **13.**

CIRCULAR INTERMITTENT MOTIONS .... Page 13
 For Axes Parallel. Locking of Driven Wheel when Idle. Avoidance of Friction Clamps. Positive Locks Classified, **13.** Work-

vii

ing Drawings of Various Circular Intermittent Motions, and Illustrations of Actual Constructions of Same, 14–18.

CHAPTER II.

SPECIAL NON-CIRCULAR WHEELS.............................. Page 19
  Complete or Incomplete. Theory of Pitch Lines. Axes Fixed. Three Cases: Axes Parallel, Axes Meeting, and Axes Crossing without Meeting. Notation. True Rolling Contact. Point of Contact on Line of Centers, 19. Radii in Pairs; their Sum a Constant. Mutually Rolled Arcs equal Each Other. Elementary Sectors. Five Special Cases. Equal Log-spirals. Will Roll Mutually and Correctly, 20. To Construct the Log-spiral Graphically. Geometric Series of Radii. Spiral Passing Two Given Points, 21. Length of Log-spiral Arc. Application of Log-spiral Wheels. Sectoral Wheels. Log-spiral Levers, 22. Weighing Scales. Wire Cutter. Wipers. Complete Wheels, 23. Symmetrical and Unsymmetrical Log-spiral Wheels: By Aid of Two Spirals; and by Angles and Radii, 24. Multilobed Wheels: by Reduction of Angles: by Assumed Angles and Limiting Radii, 25. Interchangeable Multilobed Log-spiral Wheels: as derived from One Spiral with Lobes Symmetrical, as derived from Two Spirals with Lobes Unsymmetrical, 26. Interchangeable Log-spiral Lobed Wheels. One, and Three Lobed Wheels, as determined from Proportional Sectors, 27, 28. Case of Infinite Number of Lobes, 29. Easement Curves. Equal and Similar Ellipses. Proof, 29. Range of Velocity-ratio, 30. Example from Centennial of 1876 of Elliptic Gear Wheels, 31. Interchangeable Multilobed Elliptic Wheels, *Holditch*, 32. Working Drawings and Photo-process Copy. Wheel with Infinite Number of Lobes, 33. Equal and Similar Parabolas as Rolling Wheels. Equal Hyperbolas, 35, 36. Transformed Wheels. Three Rules, 37. Changes Characteristic of Application of Each Rule, 38. Interchangeable Multilobes by Transformation, 39. Transformed Elliptic Wheels, 40. Transformed Parabolic Wheels. Wheels of Combined Transformed Sectors, 41, 42.

CHAPTER III.

NON-CIRCULAR WHEELS IN GENERAL......................... Page 44
  Three Cases: One Wheel Given to Find its Mate; Laws of Motion Given to Find the Wheels; Similar and Other Wheels from Auxiliary Sectors. Complete or Incomplete. Given One Wheel, to Find its Mate, 44. Repeated Trials. Graphic Rules for Equivalence of Lines and Arcs, 45. To Draw a Tangent to any Curve, 46. Application of Above Rules to Case of One Wheel Given to Find its Mate, 47. Examples of Above. Laws of Motion Given to Find the Wheels, 48, 49. Proof of Above, 50, 51. Photo-process Copies from Examples of Above, 52, 53. Solutions of Some Practical

CONTENTS. ix

Problems, in Laws of Motion Given to Find the Wheels. The Shaping-machine Problem, 54, 55. A Bobbin-winding Problem; Solution for Conical Bobbin, 56, 57. Motion of Driver a Variable. Case of Multilobed Wheels. Similar and Other Wheels from Auxiliary Sectors, 58. Similar and Equal Multilobed Wheels. Dissimilar Multilobed Wheels, 59. Multilobes of Unequal Number of Lobes, 60, 61.

CHAPTER IV.

SPECIAL BEVEL NON-CIRCULAR WHEELS.................................Page 62
Normal Spheres. Spherical Equiangular Spiral. One-lobed Spherical Equiangular Wheels, 62. Drawing on Spherical Blank for Pattern, 63. Two-lobed and Multilobed Bevel Non-circular Wheels. Interchangeable Bevel Non-circular Wheels, 64. Elliptic Bevel Wheels, 65. Multilobed Elliptic Bevel Wheels. Parabolic and Hyperbolic Bevel Wheels. Transformed Bevel Wheels, 66. Similar Bevel Multilobes. Dissimilar Interchangeable Bevel Multilobes. Partially Interchangeable Bevel Multilobes. Bevel Non-circular Wheels of Combined Sectors, 67. Interchangeable Multilobes with Unsymmetrical Lobes. Velocity ratio in Bevel Non-circular Wheels, 68.

CHAPTER V.

BEVEL NON-CIRCULAR WHEELS IN GENERAL....................Page 69
Laws of Motion. Plane Wheels Turned into Bevels, 69. Diagram of Plane Wheels Possessing Correct Laws of Motion. Auxiliary Diagram giving Spherical Radii from Plane-wheel Radii, 70. Spherical Blank for Gear Pattern. One or Several Lobed, 71. Proof of Above, Example, 72. Laying Out Non-circular Wheels Direct on the Normal Sphere. Case of One Wheel Given to Find its Mate. Also, Laws of Motion Given to Find the Wheels. Similar Wheels. Lobed Wheels, etc., 73–4.

CHAPTER VI.

SKEW-BEVEL NON-CIRCULAR WHEELS............................Page 75
Illustrative Hyperboloids. Pitch Surfaces of Form of Non-circular Hyperboloids, 75. First Find the Plane Non-circular Wheels of Correct Law, which Turn into Skew Bevels, 76. Spherical Blank to be Worked into a Gear Pattern. Mode of Procedure, 77–79. Proof of Above, 80.

CHAPTER VII.

NON-CIRCULAR WHEELS FOR INTERMITTENT MOTIONS..........Page 81
Classification. Cases of Axes Parallel, Meeting, or Crossing without Meeting. Easement Spurs and Locking Arcs of Greater Radius than Rolling Arcs. Easement, Locking Arcs, and Teeth in

One Plane, with Radius of Locking Arcs Less than that of the Rolling Arc, 81. Laying Out of the Pitch Curves. Photo-process Copies of Actual Bevel Wheels. Alternate Motion Gearing, 82. Limited Alternate Motions. With Velocity-ratio Constant, or Varying, 83. Unlimited Alternate Motions. Mangle Rack, 84. Mangle Wheel, 85. Axes Meeting, 86.

PART II.

TRANSMISSION OF MOTION BY SLIDING CONTACT.

CHAPTER VIII.

SLIDING CONTACT IN GENERAL. .... Page 87
Velocity-ratio in Sliding Contact, 87. Proof of Velocity-ratio, 88. Sliding and Rolling Curves with Common Law of Motion. One Sliding Curve Given, Find a Mate, With Law of Motion that of a Given Pair of Rolling Wheels, 89. Teeth of Gear Wheels, 91. Tracing of Sliding Curves. Odontoids, Centroids. Describing Templet, 92. Names, Terms, and Rules for Gear Teeth, 93. Proportions for Gear Teeth. Circumferential and Diametral Pitch, 94.

CHAPTER IX.

TOOTH CURVES FOR NON-CIRCULAR GEARING. GENERAL CASE... Page 95
Axes Parallel. Templets, 95. Use of Templets in Describing Tooth Curves, 96. Requirements of Tooth Curves, 97. The Tooth Profile. Position of Tracing Point on Describing Curve, 98. The Trochoidal Tooth Curve, 99. Form and Size of Describing Curve, 100. Individually Constructed Teeth. Non-circular Involute Gears, 101. Conjugate Teeth, 102. Limited Inclination of Tooth Curves, 103. Hooking Teeth, 104. Nearest Approach of Tooth Curve Toward Radius, 105. Practical Limit of Eccentricity of Pitch Line, 106. Eccentricity as Affected by the Tooth Profile, and the Addendum, 107. "Blocking" of the Gears, and as Affected by Length of Tooth, 108. Substitution of Pitch Line Rolling for Teeth in Extreme Eccentricity. Link Substitute for Teeth, 109. Examples and Photo-process Copies of Actual Gears, 110, 111. Internal Non-circular Gears, 111.

CHAPTER X.

TEETH OF BEVEL AND SKEW-BEVEL NON-CIRCULAR WHEELS.... Page 112
General Case of Axes Intersecting. Describing Curve a Cone, 112. Tracing of Tooth Curve Direct on Sphere. Tredgold's Approximate Method. Involute Teeth, 113. General Case of Axes Crossing without Meeting. Teeth Twisted. Correct Exact Construction not Possible 174,

CHAPTER XI.

NON-CIRCULAR INTERMITTENT MOTIONS.......................... Page 116

Teeth Spurs and Segments. Teeth as in Non-circular Wheels. Engaging and Disengaging Spurs. Backlash, 116. Spurs of Rolling, or of Sliding Curve Form. As Adapted for High Speed, or Slow, 118. Rolling Spurs often Impracticable, 119. Rolling Spurs require Excessive Arc of Motion. Bevel and Skew-bevel Wheel Spurs, 120. Solid Engaging and Disengaging Segments. Forms such as not to "Block," 121. Undercut so as to Relieve the Shock, 122. Counting Wheels. Alternate Motions. Limited in Movement, 123. Skips and Derangements. Spurs Employed, 124. Solid Segments for Reversing Motion. Best Form of Wheel for High Speed, 125. Axes Meeting. Normal Sphere, 126. Use of Motion Templets. Unlimited Alternate Motions. Mangle Wheel and Rack. Non-circular Form. Velocity-ratio, 127.

CHAPTER XII.

TEETH OF CIRCULAR GEARING.......................... Page 128

Axes Parallel. Epicycloidal, Involute, and Conjugate Gearing. Epicycloidal Gearing, 128. Some Peculiar Properties of Epicycloids and Hypocycloids. White's Parallel Motion, 129. Seven Styles of Teeth, viz.: Flanks Radial, Concave, Convex; Interchangeable Sets: Pin Gearing; Rack and Pinion; Annular Wheels. Flanks Radial. The Face and Flank Generated by Describing Circles, 130. Flanks Concave. Size of Describing Circle. Flanks Convex, 132. Interchangeable Sets. "Change Wheels." Describing Circle, 133. Pin Gearing, 134, 135. Inside Pin Gearing. Pinion of Two Teeth, 136. Rack and Pinion. Wheels of Least Crowding, 137, 138. Rack and Pinion. Flanks Radial. Flanks Concave, 139. Annular Wheels. Peculiar Case of Interference and an Extra Contact. Wheels in Sets, 140, 141. Pin Annular Gearing. Involute Gearing. Curves Described by Log-spirals Rolled on Pitch Line, 142. Practical Mode of Drawing the Teeth, 143. Rack and Pinion. Annular Wheel and Pinion. Interference of Annular Wheel Teeth, 144. Wheel of Several Styles of Teeth. Conjugate Gearing. Flanks Parallel Straight Lines, 145. Flanks Convergent Straight Lines. Flanks Circle Arcs, 146.

CHAPTER XIII.

PRACTICAL CONSIDERATIONS.......................... Page 148

Addenda and Clearance. Rule for Circumferential Pitch. Rule for Diametral Pitch, 148. To Strengthen the Teeth. Possible Clearing Curve. Path of Contact: Determined, Assumed, Limited, 149, 150. Acting Part of Plank. Line of Action. Obliquity of Line of Action, 151. The "Blocking" Tendency. Unsymmetrical

Teeth. Practical Construction of Tooth Curves, **152**. Tooth Templet, **153**. Radius Rod for Tooth Templet. To Draw Involute Teeth, **154**. Approximate Teeth in Practice, **155**. *The Templet Odontograph.* As a Ready-made Tooth Templet. Radius Rod, **156, 157**. *The Willis Odontograph,* gives Center for Circle Arc Tooth Curves, **158-160**. *The Three Point, or Grant's Odontograph,* **160, 161**. Co-ordinating the Tooth Profile, **162**. Advantages of Each of Above Instruments, **163**.

CHAPTER XIV.

MACHINE-MADE TEETH... Page **164**

Teeth with no "Hand and Eye" Process. Epicycloidal Engine to Form Epicycloidal Curve, **164**. Grinding the Tool to the Curve, **165**. Forming the Gear Cutting Tool, **166**. The Brown & Sharpe Involute Cutters; also their Epicycloidal Cutters, by Machinery of Pratt & Whitney Co., **167**. Saug's Theory of Conjugating the Teeth. The Swasey Engine with Split Cutter, **168**. The Grant's Engine with Solid Worm Cutter, **170** Form of Cutter Teeth, **171**. The Gleason's, Corliss', and Bilgram's Gear Planing Machine. Stepped and Spiral Spur Gearing, **172, 173**.

CHAPTER XV.

CIRCULAR BEVEL GEARING..................................... Page **174**

Correct and Approximate Solutions. The First by Use of Cones with Bases formed to the Normal Spheres; and the Second by Use of Normal Cones of Tredgold, **174**. Complete Drawing by Tredgold's Method, **175**. Spiral Bevel-wheel Teeth, **176**.

CHAPTER XVI.

TEETH FOR CIRCULAR SKEW-BEVEL GEAR WHEELS............. Page **177**

Approximate Construction. A Practical, Theoretically Correct Solution not Known. Approximate Epicycloidal Form. Same Arbitrarily Dressed or "Doctored," **177**. Tredgold's Method. Applied on Normal Cones, **178**. Amount of Error Illustrated, **180**. Exact Solution in *Olivier* Spiraloids, **181**. *Olivier* Spiraloid Explained. Interchangeability of Spiraloids, **182**. Interference of Spiraloid Teeth. Nature of Contact of Spiraloid Teeth, **183**. Results of an Example. Principle Demonstrated. Flat Faces of Teeth when Remote from Gorge Circles, **184**. Certain Forms Approximating the Olivier Gears. Olivier Worm and Gear, **187**. Hindley Gears. Skew-pin Gearing. Intermittent Motions, **188, 189**.

CHAPTER XVII.

CIRCULAR ALTERNATE MOTIONS............... .... Page **190**

Limited Alternate Motions. Solid Engaging and Disengaging Parts, **190**. With Attached Engaging and Disengaging Parts. The Mangle Wheel. Unlimited Alternate Motions, **191**.

CHAPTER XVIII.

CAM MOVEMENTS.. Page 192
  Cams in General. Friction. Backlash. Solution by Co-ordinates, 192. By Intersections. Use of Templets. Velocity-ratio, 193, 194. Continuously Revolving Cam, 195. Cylindric Cam, with Straight Path for Follower, 196. Same with Curved Path of Follower. Conical Cam, 197. Spherical Cam, 198. Cam Plate. Flat-footed Follower, with Specific Law of Motion of Follower. Uniform Motion, 199. Law of Motion that of Crank and Pitman. Law that of a Falling Body, 200, 201. Tarrying Points. Uniform Reciprocating Motion. The Heddle Cam, 202. Easements on Cams. Arbitrary Easements, and those Confined to Assigned Sectors, 203. Case of Flat-footed Follower, 205. Cams of Constant Diameter, 206. Cams of Constant Breadth, 207. Cams with Several Followers. The Effect of Two Followers, 208. Four-motion Cam. Illustration from Example, 210. Duangle Cam. Return of Follower by Gravity, 211. Positive Return: Three Ways, 212. To Relieve Friction, 213. Modification of Cam to Suit Form of Follower, or Roller, 214, 215. Best Form of Roller. Action of Roller upon its Pin or Shoulder. Roller for Conic Cam. For Spherical Cam, 216, 218.

CHAPTER XIX.

INVERSE CAMS AND COUPLINGS .. .. Page 219
  The Pin and Slot. Form of Slot. Velocity-ratio, 219. Oldham's Coupling, with Shafts in Parallel. 220. Shafts not Parallel, 221. Peculiar Coupling. Oldham's Three-disk Coupling, 223, 224.

CHAPTER XX.

ESCAPEMENTS.. Page 225
  Power Escapements, 225. Anchor Escapement, 226. Pin-wheel Escapement, 228. Gravity Escapement, 229. Cylinder Escapements, 231. Lever Escapement, 232. Duplex Escapement, 233. Chronometer Escapement, 234.

PART III.

BELT GEARING.

CHAPTER XXI.

TRANSMISSION OF MOTION BY BELTS AND PULLEYS. ROPE, STRAP, OR CHAIN OVER SECTORAL AND COMPLETE PULLEYS. MOTION LIMITED OR CONTINUOUS. VELOCITY-RATIO VARYING....... Page 236
  Velocity-ratio, 236. Line of Action. Non-circular and Circular Pulleys. Law of Perpendiculars given to Find the Wheels, 237.

Equalizer of Gas-meter Prover. A Draw-bridge Equalizer, **239**. Barrel and Fusee, **240**. Non-circular Pulley for Rifling Machine, **241**. Example of Treadle Movement, **243**.

CHAPTER XXII.

CIRCULAR PULLEYS.. Page **244**
Continuous Belt. Velocity-ratio. "Slip," **244**. Retaining Belt on Pulley. Pulley with High Center, **245**. Crossed Belt. Quarter-twist Belt. Guide Pulley, **246**. Any Position of Pulleys. Cone Pulleys. Problem of Cone and Counter Cone. Geometric Series of Speeds, **247, 251**. Rope Transmission: Two Systems. Rope Belting, Short and Long Stretch, **251–53**. For Haulage Lines. Compensating System. Chain and Sprocket Wheel. Teeth of Sprocket Wheels. Practical Application of Chains, **254–57**.

PART IV.

LINK-WORK.

CHAPTER XXIII.

RODS, LEVERS, BARS, ETC. THE GENERAL CASE............... Page **258**
Velocity-ratio, **258**. Peculiar Features of Link-work Mechanism. Lightest-running Mechanism. Inflexible Law of Motion. Axes Parallel, **259**. Examples: 1. Needle-bar Motion; 2. Corliss' Valve Gear; 3. Driven Piece Half the Time nearly Quiet; 4. Small Movement in a Given Time, **260, 261**. Path and Velocity of Points. Sliding Blocks and Links, **261–63**.

CHAPTER XXIV.

A ROLLING NON-CIRCULAR WHEEL EQUIVALENT FOR EVERY PIECE OF LINK-WORK. GABS AND PINS............................ Page **264**
Examples showing Several Rolling Curves, **264**. Curves and their Linkages shown Separately, **265, 266**. Rolling Curve Equivalent of Crank and Pitman. Two Cranks and Drag Link. The *Sylvester* Kite, **267, 268**. Equal Cranks in Opposite Motion. Hyperbolas. Elliptic Wheels, **269–71**. Parabolic Wheels. Unsymmetrical Link-work and Rolling Wheels, **272, 273**. Dead Points in Link-work. "Gab and Pin." Path of Gab and Pin. Gab and Pin in Practice, **274–77**. Multiplying Motion by Link-work, **278**.

CHAPTER XXV.

CONIC LINK-WORK .... Page **279**
Axes All Meet in a Point. Cranks Equal or not. Velocity-ratio, **279**. Rolling-wheel Equivalent of Conic Link-work. Dead Center, and Gab and Pin in Conic Link-work. Examples of Peculiar Move-

ments, **281**. Possible Valve Motion. Hooke's Joint or Coupling, **282–85**. Almond's Coupling, **286**. Crank Coupling, **287**. Reuleaux Coupling, **288**.

CHAPTER XXVI.

LINK-WORK WITH AXES CROSSING WITHOUT MEETING.......... Page 289
    Ball and Socket or Sliding Joints. Velocity-ratio, **289**. The *Willis* Joint System, **290**. Shafts Connected by Cranks and Angle Blocks, Three Examples, **291**, **292**. Ratchet and Click Movements. Running Ratchet. Varied Step Movement, **292**, **293**. Reversible Running Ratchet. Reversible Varied Rate Running Ratchet. Continuous Running Ratchet. Forms of Teeth and Click, **294**, **295**. Friction Ratchets. Fluted Bearing. Continuous Friction Ratchet. Wire Feed Ratchet. Running face Ratchet and Click, **296**, **297**.

PART V.

REDUPLICATION.

CHAPTER XXVII.

PULLEYS OR SHEAVES AND ROPES............................ Page 298
    Velocity-ratio. Parallel Ropes, **298**. Ropes not Parallel. Velocity-ratio. Hay Unloader, **299**. Hydraulic Elevators. Weston Differential Block. Velocity-ratio, **300**.

PRINCIPLES OF MECHANISM.

INTRODUCTION.

In working out the design, drawings, and specifications for a machine, the form, strength, and motion of the various parts must be determined, the last being the object of the *Principles of Mechanism*.

In Principles of Mechanism we find the application to machines, of the principles of Kinematics, or Cinematics, the elementary combinations of mechanism of which machines, being studied separately.

A *Machine* may be defined as a combination of fixed and moving parts or devices, so disposed and connected as to transmit or modify force and motion for securing some useful result.

The fixed parts constitute the frame or supports for the moving parts.

The moving parts constitute a train or trains of mechanism.

A train of mechanism may be primary or secondary; the former being supported directly by the frame, and the latter by other moving parts.

All the moving parts of machines may be regarded as mechanical devices and classified as follows:

1st. Revolving shafts.

2d. Revolving wheels or cams, with or without teeth.

3d. Rods or bars with reciprocating or vibratory motion, or both.

4th. Flexible connectors depending on friction.

5th. Flexible connectors independent of friction.

6th. A column of fluid in a pipe.

Trains of mechanism consist of combinations of the above devices, the least number securing a modification of force or motion in a given case, being an elementary combination.

The study of the *motions* of a *machine* is usually pursued by taking up separately the *elementary combinations* of mechanism composing the machine.

Professor Robert Willis of Cambridge, England, was the first to present a thorough, systematic, and comprehensive table of the elementary combinations of all mechanism. In our study of mechanism to include all kinds and varieties without omissions, we can do no better than to follow this table, as below:

SYNOPTICAL TABLE OF THE ELEMENTARY COMBINATIONS OF MECHANISM. (WILLIS.)

Mode of Transmission of Motion.	Directional Relation Constant.		Directional Relation Changing Periodically.
	Velocity-ratio Constant.	Velocity-ratio Varying.	Velocity-ratio Constant or Varying.
By rolling contact.	Rolling cylinders, cones, and hyperboloids; pitch-circles of circular gear-wheels and sectors.	Pitch-lines of non-circular gearing, complete or sectoral.	Pitch-lines for mangle-wheels and mangle-racks; limited alternate motions.
By sliding contact.	Tooth-curves; segmental cams; screws; worm gearing.	Tooth-curves for noncircular gearing; cams; pin and slotted lever; irregular worm gearing; stop motions.	Cams in general; pin and slotted lever; double screw; swash-plate; escapements.
By wrapping connectors, or belt gearing.	Band or belt and pulleys; chain and sprocket-wheel.	Cam-shaped pulley and belt; fusee and chain or chord.	Cam-shaped pulley and lever, or tightener; treadle motion.
By link-work.	Equal cranks with link; lever with proportional links; bell-crank and links.	Equal cranks with link; unequal cranks with link; Hooke's universal joint.	Crank or eccentric and pitman; ratchet-wheels and clicks; unequal cranks and link.
By **reduplication**.	Cord and pulley; pulley-blocks with parallel ropes or chains.	Pulleys, with rope or chain not parallel.	

NAMES AND TERMS.

In the study of mechanism certain terms are used, some of which are defined below:

Velocity.—Time is required for a point to move a distance along a line or path. For *uniform motion* the distance moved over per unit of time is the velocity, usually considered as *feet per second*.

Thus velocity is the rate of motion, or movement.

For the case of uniform motion, any distance passed over in a corresponding given time, divided by that time, gives the velocity.

When the velocity is not constant, the space passed over in a very short time is divided by that short time; as, for instance 0.3 ft. in 0.1 sec., when the velocity will be

$$\frac{0.3}{0.1} = 3 \text{ ft. per sec.},$$

which is the velocity or rate of moving at the instant considered.

In mathematical calculations the space and time are often reduced to infinitesimals for extreme exactness, but this is rarely necessary in the study of mechanism.

Angular Velocity.—Velocity may here be distinguished as linear and angular; the former being the rate of motion along any line, while the latter is the *rate of motion of a point at a distance unity from a center of angular motion*. Either may be constant or variable. If variable, the rate is to be expressed for an instant by taking the ratio of an element of space to the corresponding element of time.

Velocity-ratio.—This is the ratio of two angular or two linear velocities; or, in some cases, of an angular and a linear velocity.

Period.—Moving parts of machines continue in the repetition of certain definite complete movements.

The time for one of these complete movements may be called a *period*, though this term is but little used in mechanism.

Cycle.—The time for several moving pieces, considered collectively, to return to their given initial positions may be called a *cycle*, though this term is of comparatively little use in mechanical movements.

Revolution.—A revolution is to be considered a complete turn, while the term *rotation* may be applied to any portion of a turn.

Driver and Follower.—In any elementary combination of mechanism one piece always drives the other, the one therefore being called the *driver*, and the other the *follower*.

Directional Relation.—This term has reference to the relation of the directions of motion of driver and follower. When one never reverses its direction of motion unless the other does also, the *directional relation is constant*, otherwise it is said to be variable.

PART I.

TRANSMISSION OF MOTION BY ROLLING CONTACT.

CHAPTER I.

ROLLING CONTACT IN GENERAL.

Rolling Contact in elementary combinations of mechanism implies a driver which turns a follower by rolling against it without slipping, the driver and follower *having axes of motion* which are at a *constant distance apart*.

The **Line of Contact** of thick pieces is always a straight line like that of the contact of a pair of parallel straight cylinders tangent to each other.

Sometimes reference is had to the *plane of the axes*, or plane which contains both axes.

The line of contact of thick pieces in true rolling contact will evidently remain in the plane of the axes, and in transverse sections the *point of contact* or of tangency of the rolling curves of section must likewise remain on the line joining the centers of motion, called *line of centers*.

It is evident that in true rolling contact no slipping can occur, and that arcs which have rolled correctly in mutual contact must be equal to each other.

VELOCITY-RATIO IN ROLLING CONTACT.

First. *For Circular Arcs and Cylinders.*—Suppose Fig. 1 to represent a transverse section of a pair of rolling circular cylinders with axes at A and B. The line of centers is AB, and C on that line is the point of contact. Take $Cb = Ca$, and draw the radii R and r. Now if Ca rolls without slipping on Cb, a and b

will eventually come into contact at C on the line of centers, when R will have the position AC, and r the position BC; while the axis A will have turned through the angle CAa, and the axis B through the angle CBb. The point e will describe the arc ed, and the point g the arc gf. If these movements occur in one

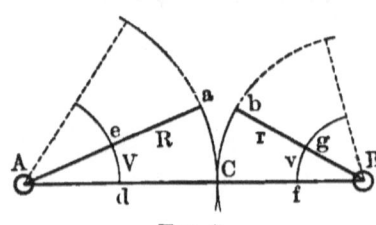

FIG. 1.

second of time with uniform motion, and if e and f are taken at a units distance from A and B respectively, then

ed = angular velocity of $A = V$;
and gf = angular velocity of $B = v$.
Also $Ae : R :: de : Ca :: V : Ca$,
$Bg : r :: fg : Cb :: v : Cb$ or Ca;
and $V \times R = Ca \times Ae = v \times r = Cb \times Bg$.
Whence $\dfrac{V}{v} = \dfrac{r}{R} = \dfrac{BC}{AC}$,
since $Ca = Cb$, and $Ae = 1 = Bg$,

which is the *velocity ratio* of the axes A and B for motion transmitted from A to B by the arc Ca rolling on the arc Cb, and equals the inverse ratio of the radii R and r.

As this is true of any transverse section of the cylinders, it is true of the entire cylinders. Hence the important principle, viz.: *the velocity-ratio of truly rolling cylinders is equal to the inverse ratio of the radii of those cylinders.*

Second. *For Non-circular Arcs and Cylinders.*—Let Fig. 2 represent a transverse section of a pair of correctly rolling non-circular cylinders, or portions of them, with A and B as axes. Then AB is the line of centers, C the point of contact, aAc and bBd the non-circular sectors. Draw circular arcs eCg and fCh, and the velocities V and v as shown. Then V is common to aAc and eAg. Also v is common to bBd and fBh, and the velocity-ratio is the same for the non-circular as for the circular sectors, the angles to the sectors being regarded as very small.

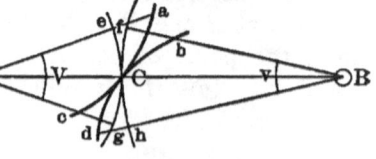

FIG. 2.

The segments of the line of centers being the same for the non-circular as for the circular sectors, we have

$$\text{Velocity-ratio} = \frac{V}{v} = \frac{BC}{AC};$$

that is: *the velocity-ratio in non-circular wheels at any instant is equal to the inverse ratio of the segments of the line of centers.*

As these segments are continually varying in non-circular wheels, it follows that the velocity-ratio for such wheels is also variable.

CIRCULAR ROLLING WHEELS.

DIRECTIONAL RELATION CONSTANT. VELOCITY-RATIO CONSTANT.

Friction Wheels and Pitch Lines for Circular Gearing.

For this subject the wheels are circular, with velocity-ratio constant; and we have three general cases.

I. **Axes Parallel.** For this case there are two subdivisions. First: Contact between the axes as in Fig. 3. Second: Contact outside the axes as in Fig. 4.

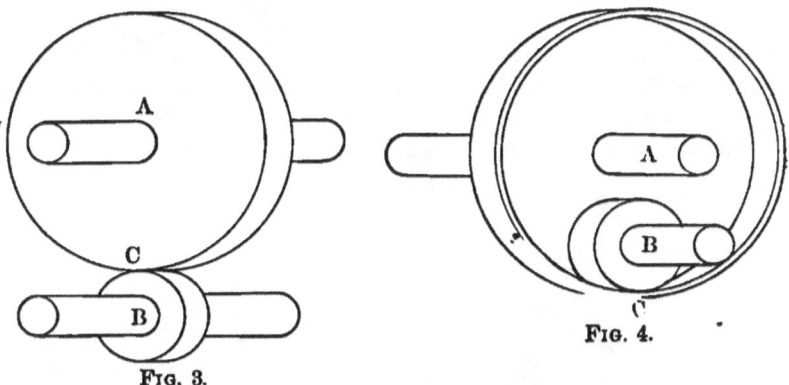

Fig. 3.   Fig. 4.

These *friction wheels*, or *pitch lines*, as the case may be, are so simple in the theory of mechanism that no further treatment seems necessary, unless the following graphic problems are considered.

First. *Contact Between the Axes.*—The axes being at a fixed distance apart and parallel, let them be represented in Fig. 5 by the lines at A and B. Take the velocity-ratio as being given with the value 2/5. Then

$$\frac{V}{v} = \frac{2}{5} = \frac{r}{R}.$$

As the sum of the radii equals the distance between the axes,

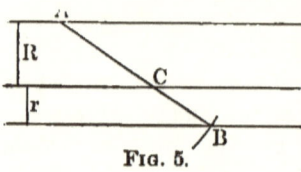

Fig. 5.

then if we take $r = 2$ inches and $R = 5$ inches, and add, we have 7 inches. If this be less than the distance between the axes, double each and add, or triple each and add, until the sum is greater than the distance between the axes, as AB, Fig. 5. Then with AB as a radius in the dividers place one point at A and strike the arc intersecting at B. Draw a straight line AB, and lay off on this line from A the greater part of the sum AC, and from B the lesser part, giving the point C. Through C draw the line of contact. Then the perpendicular distances from the axes A and B upon the line of contact will be the radii of the wheels, which will be in simple proportion with the distances BC and AC.

Second. *Contact Outside the Axes.*—Here also we take A and B for the axes, and C the line of contact, the three lines being parallel.

With the same example of velocity-ratio as before, take the difference of the values 5 and 2, or 3; or twice or more times these values if 3 is too small. Then with the dividers opened to the difference found,

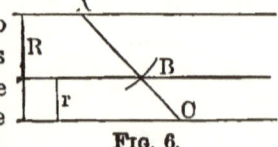

Fig. 6.

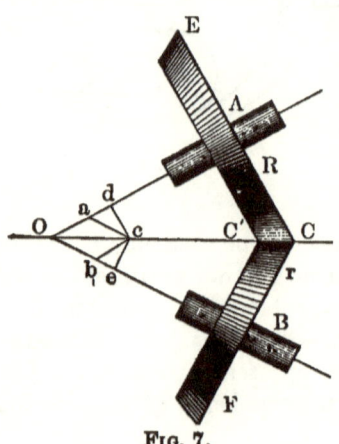

Fig. 7.

place one point of the dividers at A and strike an arc intersecting at B, and through AB draw the straight line AC, extended as far as necessary. Then with B as a center and the lesser of the above values lay off BC; and the greater should equal AC. Now drawing the line of contact through C, parallel to A and B, we have the radii as perpendiculars from A upon C, and from B upon C.

II. **Axes Meeting.** Take AO and BO as the axes, meeting at O. The diameter of the wheel A may evidently be any line CAE, perpendicular to its axis AO, and the radius $AC = R$. Then the diameter of the mating wheel must be CF, and the radius $BC = r$.

To determine the relation between the radii of the wheels and their angular velocities, lay off the velocity of A on the axis of A, and the velocity of B on the axis of B, as Oa and Ob respectively, and complete the parallelogram $Oacb$ by drawing ac and bc. Also draw cd and ce perpendicular to the axes, and we have the triangle acd similar to bce, the angle cad being equal to cbe. Then

$$V : v :: Oa : Ob :: bc : ac :: ce : dc :: r : R.$$

Whence the velocity-ratio

$$\frac{V}{v} = \frac{r}{R};$$

proving the correctness of the parallelogram for conveniently locating the line of contact when the velocity-ratio and angle between the axes are given.

If another pair of circles tangent to each other be drawn on the axes A and B, as, for instance, those tangent at the point C' on the line OC, the ratio of the radii of this pair will be the same as that of the circles tangent at C, as is plain from the geometry of the figure.

Hence correctly rolling surfaces for this case must be cones with their vertices in common as at O.

In the above the contact is between the axes, which are situated to intersect at an angle of less than 90 degrees.

When the angle between the axes is greater than 90 degrees there seems to be a second case, as if the contact were outside this angle as shown in Fig. 8. But the principles are all the same, and the wheels are readily made in practice.

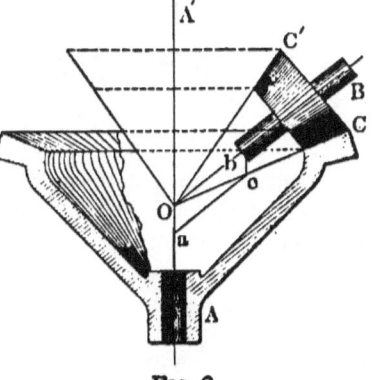

Fig. 8.

The wheel A of Fig. 7 is shown at A', Fig. 8, so that the cone B may have a mate AC or $A'C'$ on the axis A.

A peculiar result is obtained when the line OC is perpendicular to AO, the wheel A being a plane circular disk instead of a cone. But O, the intersection of the axes, is the common vertex of all the rolling surfaces $AA'B$, etc., in every instance of correct rolling contact.

10 PRINCIPLES OF MECHANISM.

Some old mill gearing made wholly of wood ignores the above conic forms of the theoretic surfaces, where one wheel has teeth upon its side and the other upon its edge.

III. Axes Crossing Without Meeting.—For convenience, take the vertical projections of the axes and the line of contact of the rolling surfaces as here shown, parallel to the "ground line" of the figure This line of contact must be a straight line, and an element of each of the surfaces.

If the surfaces have a possible existence for this peculiar case, it is plain that the form of each may be conceived to be described

FIG. 9.

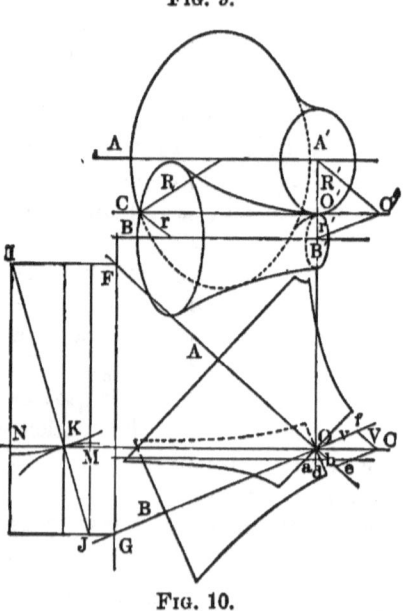

FIG. 10.

by imagining the line of contact to be fixed to the one axis and to be revolved about it, thus generating that surface, first for the one, and then for the other. The form of surface thus described is the *hyperboloid of revolution* with axes AA' and BB', Fig. 9. The smallest diameters will be at the common perpendicular, or shortest distance between the axes $A'B'$, which perpendicular is intersected at O' by the line of contact $C'O'C$.

To locate the line of contact in the horizontal projection, imagine the surfaces to be extended to infinity, where they become rolling cones. Thus we can locate the line of contact in Fig. 10 to correspond with the velocities V and v as in conic wheels, Fig. 7, laying off V on the axis of A, and v on the axis of B, as shown, and completing the parallelogram with CC, the diagonal on the line of contact in Fig. 10 when sufficiently extended. In the drawing, it is most convenient to place the line of contact CON parallel to the ground line, as in Fig. 10.

To locate the line of contact in Fig. 9: First, it is plain that it will be parallel to the axes in order to meet the case of rolling cones of infinite extent. Second, to find the point O' which divides the common perpendicular $A'B'$ into segments, or gorge-circle radii, draw $A'C'$ and $B'C'$ parallel to the respective axes A and B of Fig. 10. Through the intersection C', Fig. 9, and parallel to the axes draw the line of contact. Thus all the essential lines of Figs. 9 and 10 become located, and the hyperboloids can be drawn in as shown, using the axes A and B, Fig. 10, as geometric axes, and the line of contact OCN as an asymptote for the hyperbolic curves. Afterwards Fig. 9 can be completed by the theory of projections.

The correctness of this construction for the triangle $A'B'C'$ is shown by aid of the end view at the left of Fig. 10, obtained by passing a plane, FG, perpendicular to the line of contact, and revolving it to $IJMN$, in which some point K is where the line of contact pierces this plane. In this cross-section, the point K is the point of contact of the rolling hyperboloidal surfaces as intersected by the plane, in curves tangent at K, as shown. As I and J are centers of rolling motion in this plane, with IK and JK as radii, K must be on a straight line from I to J, which locates the line of contact as between N and M, or between A' and B' as well; since the line of contact KC, $O'C'$ is parallel to the projection of the axes in Fig. 9. But the triangle $A'B'C'$ is similar to FGO, and INK to JMK, so that $A'O' \div B'O' = NK \div MK$, proving true the construction Fig. 9 for the position of $C'O'$.

It may be noted that a system of hyperboloids adapted for rolling upon each other in pairs interchangeably, may be drawn by making K, CO the line of contact, O the point of intersection of all the axes in plan, IJ the locus of all their intersections with the plane FG, all axes being parallel in their projections Fig. 9.

A notable point in the history of this case is that the early writers, from analogy with the case of bevel wheels, assumed that the radii R' and r' were in the simple inverse ratio of the angular velocities. The French author Belanger, however, gave an extended and rigorous proof to the contrary, agreeing with the above graphical construction $A'B'C'$ of Professor Rankine. To show this construction to be correct, suppose the wheels to make a small turn in rolling contact so that the line of contact, OC produced, makes a small displacement from OC to ab, Fig. 10. Then Oa may be regarded as the horizontal projection of the corresponding movement of the point O' in the circle $A'O'$. Likewise Ob may be the plan of the corresponding movement of O' in the circle $B'O'$. But a and b should be in a line parallel to CO, as required for no *lateral* slipping of the surfaces; longitudinal slipping being necessarily allowed to provide for the endwise sliding of the teeth of these wheels while engaged, and in action, the teeth being laid out to coincide, in position, with a series of lines of contact, or straight elements of surface properly distributed around the hyperboloids as explained under *Teeth of Wheels*.

Thus conditioned, we see that the triangle Oab is similar to $A'B'C'$.

To prove this construction correct we have from Fig. 10

$$R'V = Oa, \quad r'v = Ob, \quad Oa \cos aOd = Od = Ob \cos bOd,$$

and from the parallelogram OC we have

$$V \sin eOC = V \sin aOd = v \sin fOC = v \sin bOd,$$

whence

$$\frac{R'}{r'} = \frac{v \cdot Oa}{V \cdot Ob} = \frac{v \cos bOd}{V \cos aOd} = \frac{\sin aOd}{\sin bOd} \times \frac{\cos bOd}{\cos aOd} = \frac{\tan aOd}{\tan bOd}$$

a relation that must exist between R', r', v, and V, as consistent with the allowed longitudinal sliding along ab, but not lateral slipping of surfaces.

From Fig. 9, as constructed, we obtain

$$O'C' \tan A'C'O' = R', \quad O'C' \tan B'C'O' = r',$$

whence

$$\frac{R'}{r'} = \frac{\tan A'C'O'}{\tan B'C'O'} = \frac{\tan aOd}{\tan bOd}$$

the same relation as obtained from Fig. 10 as necessarily constructed. Hence the construction for $A'O'B'$ as described is the correct one.

These wheels, sometimes called *rolling hyperboloids*, can never serve as practical rolling surfaces or friction wheels, from the fact that the unavoidable longitudinal sliding induces lateral slip; but in skew-bevel gearing they are entirely practical. To determine the longitudinal slip, compare ab with Oa or Ob, these being corresponding values of this slip, the movement of the end of R'', or of r', respectively.

INTERMITTENT MOTIONS.

These might have been classified with respect to the relation of axes, as parallel; meeting; or not parallel and not meeting; but it is believed advisable to classify them with regard to their peculiarities, and treat them here together.

This name is given to movements in rolling contact where the driver A, Fig. 11, moves continuously while the follower, or driven piece B, has periods of rest and motion.

The class here considered, of velocity-ratio constant, includes such of these movements as have circular rolling arcs, here dotted as pitch lines for the teeth.

To be thoroughly practical, the driven piece should be positively locked when idle, and be stopped and started in gradation of motion to avoid shocks and breakage.

Spring checks, friction clamps, etc., should never be relied upon, as a momentary change in speed may prevent the spring from properly catching in its notch, or the friction clamp from wear may get out of adjustment and fail to act. Serious breakages have been known to occur in machines for making wire hooks for hooks and eyes, where this movement has been used with a friction clamp instead of positive locks.

These movements possessing positive locks are made in at least two ways which are thoroughly practicable.

First, with the locking arc of greater radius than the rolling arc, and provided with easement spurs.

Second, with the locking arc of smaller radius than the rolling arc, and provided with easement segments.

First. Here the locking arc JK on the wheel A, Figs. 11 and 12, is given a larger radius than the rolling arc DHE, in order that the curves, EJ and DK, of varying radius may act upon the saddle curve FG while the first point F or the rolling arc of B is moving into engagement with the first point of the rolling arc EH, to prevent the wheel B from moving out of place in the one direction, the spurs *L* and *N* preventing derangement in the other direction. The spurs *L* and *N* are on the front side of the wheels, while the spurs *P* and *R* are on the opposite sides, and dotted.

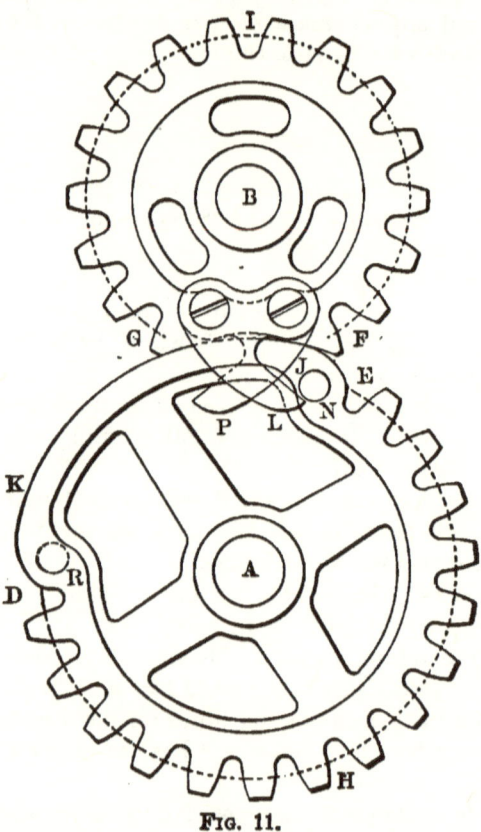

Fig. 11.

In Fig. 11 the pin *N* moves toward the spur *L*, fastened upon *B*, striking it first as the locking arc *JGK* begins to fall away, the latter being cut away more and more toward *E* so that *B* is just allowed to turn with but slight backlash until the pin acting against the spur has thus brought the initial point of the rolling arc *FI* to contact with the initial point of the rolling arc *EH*, when rolling action begins and the spurs *L* and *N* go out of use for that revolution. When the rolling has continued through *H* to *D*, the locking arcs and spurs again come into action until *B* is locked again in the position shown, it having made one revolution while *A* made but part of one, *B* remaining idle and locked while *A* completes its revolution.

It will be seen that the pin strikes the spur quite abruptly in Fig. 11 unless the spur is considerably inclined, and starts *B* sud-

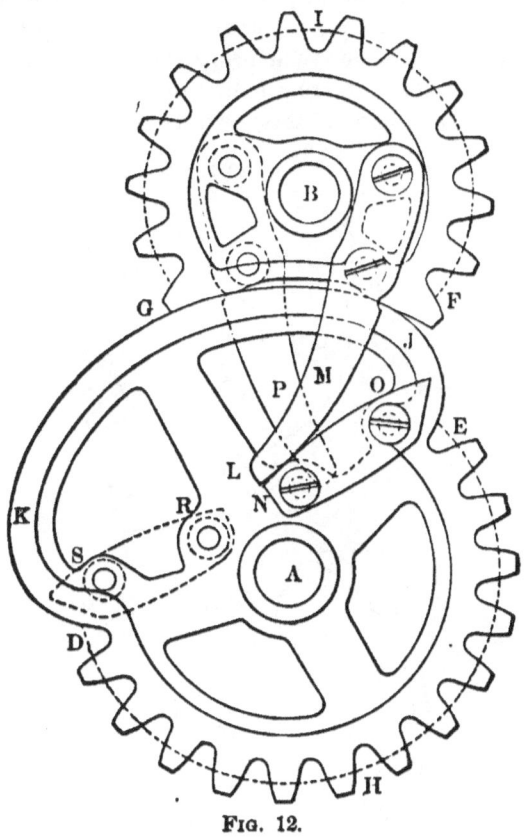

Fig. 12.

denly, so that in high speed the pin or spur may be broken, while in Fig. 12 the two spurs, *LM* and *NO*, extending nearly to the axis of motion of the driver *A*, start *B* slowly at first and with greatly reduced shock, but as the contact between the spurs moves outward, the wheel *B* is accelerated until *F* meets *E* when rolling contact begins. By placing a pair of spurs on each side of the wheels, they are adapted for running either way; or if in one direction only, it is important to have one pair for accelerating *B* at starting, and another for retarding it gradually while approaching

16　PRINCIPLES OF MECHANISM.

the locking arc. Thus the length of the spurs may be fixed in a particular case by the judgment of the designer.

The rolling-circle arcs of these wheels are what bring them under the present topic of rolling contact and constant velocity-ratio. The spurs will usually work by sliding contact, though they may act by rolling, but with velocity-ratio varying, and their forms may be studied under their proper headings. (See Fig. 141.)

Second: with locking arc smaller in radius than the rolling arc, as in Fig. 13, where B is shown as approaching the locked

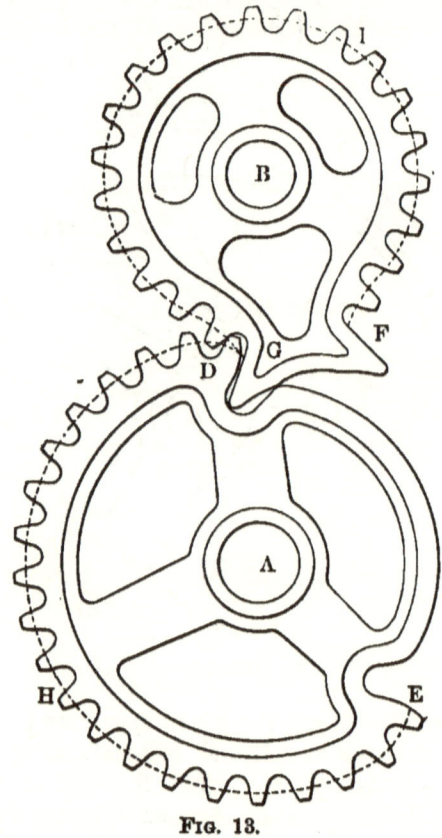

Fig. 13.

position, and with H and I the rolling arcs. As E approaches F it glances and slides on F in such a way as to start B slowly, and with acceleration until the initial points of the rolling arcs H and I come to contact when rolling begins with velocity-ratio constant through the rolling circular arcs. When D approaches G, the reverse action occurs until B is again locked on the locking arc as shown to soon occur.

In the details of these wheels, the locking arcs are cut away to some extent, as shown near E and D, to allow the points F and G to turn while passing from lock into action, and B is cut away somewhat in the lock between F and G. Also the circular rolling arcs H and I have easements at their ends, thus introducing short non-circular rolling arcs at E and D, to be set with gear teeth, when they are laid out for the rolling arcs treated as "pitch lines" for teeth.

Fig. 13 has the advantage of simplicity, A and B being each in practice cast in one piece, while Figs. 11 and 12 require spurs, pins, etc., which usually must be made and put on with screws or

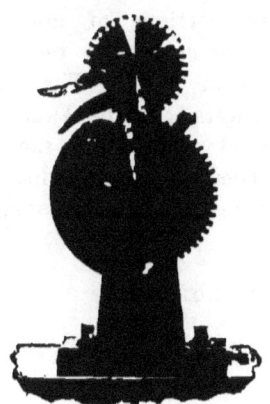

FIG. 14.

FIG. 15.

rivets. The form Fig. 13 is used on certain binders of reaping machines.

In these figures there is but one locking arc, or lock, to each wheel, so that A makes as many revolutions in a given time as B. But it is readily seen that A may have several rolling and locking arcs in alternation in a circumference; likewise B may have several rolling arcs and locks in its circumference, and the number differing from that of A, as, for instance, A may have three and B two.

These wheels for intermittent motion may be made with *axes meeting*, or as *not parallel nor meeting*, under the principles for those cases as already laid down except for the easement spurs, and segments, which, whether acting by rolling or sliding contact, may be treated by principles considered later, which see.

In practice, the rolling arcs in the above intermittent wheels must be set with gear teeth as above mentioned, the proper construction of which teeth will be considered later under Teeth of Wheels.

To give a clearer idea of these wheels when complete with teeth, locks, spurs, easements, etc., photo-process copies are here given, where Fig. 14 is to represent Fig. 12 above, as complete and ready for action. The rolling arcs of Fig. 12 are here provided with teeth.

Fig. 16.

In Fig. 15 is illustrated a pair of wheels explained in Fig. 13, where the circular rolling arcs are terminated with short portions of non-circular arcs or easements, the teeth being extended over both. The distance between centers is about $4\frac{1}{2}$ inches.

In Fig. 16 we have the same as in Fig. 15 turned into bevel wheels with axes meeting at an angle of 30 degrees, though the angle may be 90 degrees more or less.

CIRCULAR ALTERNATE MOTIONS.

The few examples for this case are deferred to page 82.

CHAPTER II.

ROLLING CONTACT OF NON-CIRCULAR WHEELS.

VELOCITY RATIO VARYING.—PITCH LINES OF NON-CIRCULAR GEARS.

These may be *complete*, that is, filling the full circle of 360 degrees, so as to admit of continued rotation, one revolution after another, or they may be incomplete or sectoral.

This subject may very properly be regarded as the general theory of pitch lines of gearing; and as it is a subject of much importance it will be treated at some length.

The axes of motion are supposed to be fixed in position, and, as in circular wheels, there are three cases, viz.: axes parallel; axes meeting; and axes not parallel and not meeting.

Case I. Axes Parallel.

Here, as in circular wheels, the contact point may be between or outside the axes.

The following notation will be convenient for use:

$A =$ the driving wheel, or axis.
$B =$ the driven wheel, or axis.
$C =$ the point of contact of wheels.
$c =$ the distance between axes.
$R =$ the radius of wheel A, and $= AC$.
$r =$ the radius of wheel B, and $= BC$.
$V =$ angular velocity of wheel A.
$v =$ angular velocity of wheel B.
$S =$ stated, or definite, portion of arc of A.
$s =$ stated, or definite, portion of arc of B.

The essential conditions for true rolling contact of these **wheels** are:

First. The axes must be fixed at a distance c apart.

Second. The point of contact must remain on the line of centers.

Third. The sum of any *pair* of radii must equal a constant,

or
$$R + r = c = \text{const.}$$

Fourth. Arcs having rolled truly in mutual contact are equal,

or
$$S = s.$$

For convenience in the study of these wheels, both A and B may be regarded as consisting of elementary or *small sectors* and be treated in mating pairs, with limiting radii in pairs, and arcs in pairs, etc. Thus a pair of radii will always be $R + r = c$, and a pair of arcs will be $S = s$, the capitals belonging to driver A, and the small letters to driven wheel B.

Special Cases of Non-Circular Wheels.—There are but five simple and readily demonstrated cases of correct-working non-circular wheels, viz.:

First. *A Pair of Logarithmic Spirals of the Same Obliquity.* (*Sometimes called Equiangular Spirals.*)
Second. *A Pair of Equal and Similar Ellipses.*
Third. *A Pair of Equal and Similar Parabolas.*
Fourth. *A Pair of Equal and Similar Hyperbolas.*
Fifth. *Transformed Wheels.*

First. Equal Logarithmic Spirals.

The leading characteristic of the logarithmic spiral is that it has a *constant obliquity* throughout.

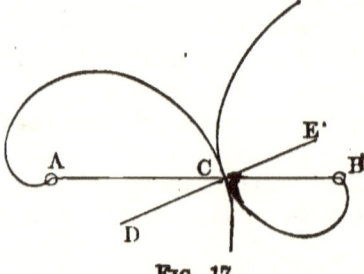

Fig. 17.

Taking A, Fig. 17, as the origin, or *pole*, of the spiral A, the angle ACD, between the radius vector AC and normal CD, is called the obliquity of the spiral, its value being the same for all points of this spiral. If a second spiral, B, of the same obliquity be placed in contact with the first, as shown in Fig. 17, the angle BCE being equal to the angle ACD, the contact point C will be found to be situated on the straight line AB, joining the poles, for whatever point of contact between the spirals.

Then taking the points, or poles A and B as fixed axes of motion, the spirals will roll upon each other without slipping, C

being the *point of contact* and AB the *line of centers*, and the conditions for rolling contact of non-circular wheels are satisfied.

TO CONSTRUCT THE LOGARITHMIC SPIRAL.

In the logarithmic spiral, for equal angles about the pole, the radii vectors are in *geometric progression*. Draw two straight lines

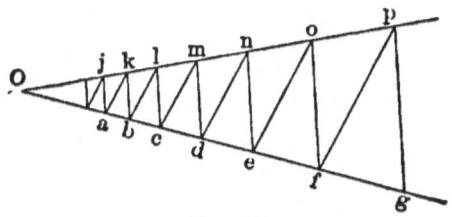

FIG. 18.

from O, Fig. 18, and a line ak and kb. Then draw bl parallel to ak, lc parallel to kb, cm parallel to lb, etc., as far as desired. Then it is found that the lines Oa, Ob, Oc, Od, etc., are in *geometrical progression*, or form a *geometric series*. The same is true of Oj, Ok, Ol, etc.; or of aj, bk, cl, etc. Hence any one of these systems of lines may be used for radii vectores of the spiral. Through O, Fig. 19, draw a series of lines at equal intervening angles, and lay off any one of the above series of lines as Oa, Ob, Oc, etc., and trace the curve abc, etc., through the points thus determined. This curve is a logarithmic spiral, the construction making it evident that the obliquity is constant.

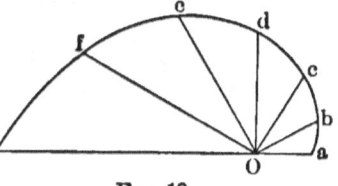

FIG. 19.

It is difficult to construct a spiral, as above, that will pass through two given located points as a and c in Fig 20. For this case draw a semicircle to a diameter $Aa + Ac$ as in Fig. 21, and the perpendicular Ab will be the radius which bisects the angle aAc, Fig. 20, thus determining the point b in the curve abc. Then a similar construction of Fig. 21 for the sector aAb will give the radius Ad; and another construction

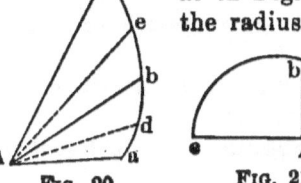

FIG. 20.   FIG. 21.

the radius *ae*, etc., in continual bisection of the angles until the desired number of points in the spiral *ac* is obtained.

LENGTH OF THE LOG-SPIRAL ARC.

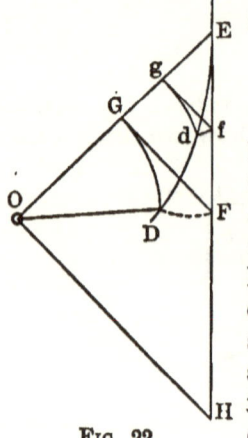

Fig. 22.

Taking O as the pole, OD and OE as the limiting radii vectores of the log-spiral sector ODE, draw a tangent EF at E. Make $OG = OD$, and draw GF perpendicular to OE. Then $EF =$ the arc DE of the spiral. For a small portion dEg this appears evident, as dE and fE are equal; and the same at the limit of every short arc. From this it appears that $EH =$ the arc of the spiral from the pole O, to E, OH being perpendicular to OE.

APPLICATIONS.

SECTORAL LOG-SPIRAL WHEELS.

Fig. 23 shows a mating pair of log-spiral sectoral wheels, where the arcs and also the differences of the limiting radii of the sectors are equal; AB being equal the sum of a *pair* of radii.

The velocity-ratio for the position of the wheels shown in the figure is $\dfrac{V}{v} = \dfrac{BC}{AC} = \dfrac{r}{R}$,

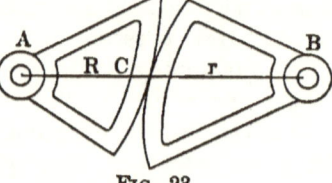

Fig. 23.

R and r being the pair of radii to the point C.

LOG-SPIRAL LEVERS.

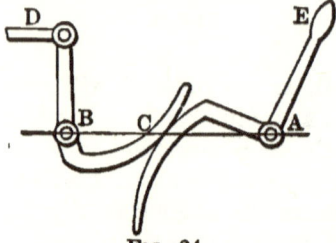

Fig. 24.

Fig. 24 illustrates the application of log-spiral levers, where a rod D is thrust forward or back by moving handle E, the levers working by rolling at C.

Another application is shown in Fig. 25 to a wire-cutter, which in practice has been found to work well.

A possible application to weighing-scales is illustrated in Fig. 26.

In Fig. 27 is shown a pair of "wipers" in steam-engine valve-

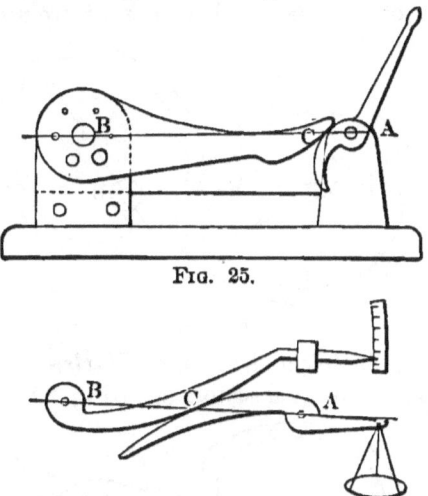

FIG. 25.

FIG. 26.

gears of the log-spiral order. Here the center B is at an infinite distance away because the driven spiral moves in a straight line, being supported on a rod sliding in guide-bearings DE. The spiral B in this case becomes a straight line as shown.

The velocity-ratio is that of the linear velocity of sliding of B, and of the angular velocity of A, or V; and we have linear velocity of $B = AC \cdot V$, or the

velocity-ratio $= \dfrac{\text{linear velocity of } B}{V} = AC$.

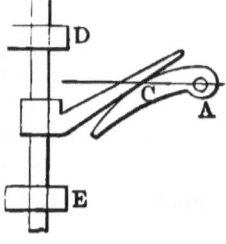

FIG. 27.

We note that the velocity of B is the same as that of the point C as revolving about A; since linear velocity of $B = AC \times V$.

COMPLETE LOG-SPIRAL WHEELS.

First. *Symmetrical Unilobed Wheels.*—In this case each wheel is composed of two equal sectors of the spiral, each sector being one of 180 degrees, as shown. These admit of continuous rotation of A and B with a variable velocity-ratio, for which, at any instant,

$$\frac{V}{v} = \frac{BC}{AC}.$$

If the velocity of A is constant, that of B will be variable.

Wheels like these are termed *lobed wheels*, because extended out to one side in a lobe. The wheels in Fig. 28 are *unilobed*.

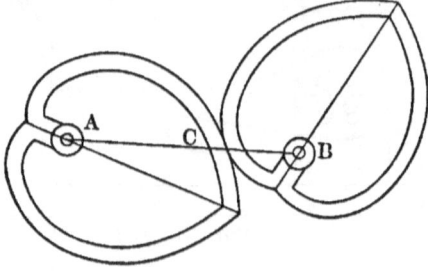

FIG. 28.

Second. *Unsymmetrical Unilobed Wheels.*—Here each wheel consists of a pair of unequal sectors as in Fig. 29.

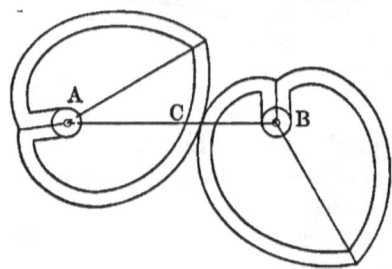

FIG. 29.

1. *By aid of two log-spirals.* In this case construct one log-spiral as in Fig. 19, say right-handed, and a second one, left-handed, on a transparent paper. By placing one over the other, with poles coincident, one may be turned around on the other until the desired size is seen lying between two consecutive intersections, as in Fig. 30. Thus ADE may be chosen, or AEF. If the first is too small and the second too large the transparent tracing may be turned to another position over the other, changing the wheels to the desired size.

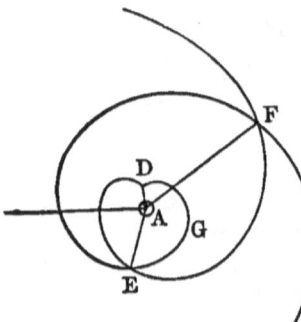

FIG. 30.

2. *By assumed angles and radii.* Here proceed as by the process of Figs. 20 and 21, where the radius Ab divides the angle aAc and is made a mean proportional between Aa

and Ac, as in Fig. 21. Again, find a mean proportional between Aa and Ab for a radius equally dividing the angle aAb, etc., etc., for as many points in ab as desired. Likewise for the arc bc.

Third. *Multilobed Wheels.*—1. *By reduction of angles.* Suppose a series of equidistant radii be drawn in Fig. 28 or 29, and that the angles thus found be reduced by one half. Fig. 29 would then be changed to Fig. 31, giving the half of a 2-lobed wheel; the

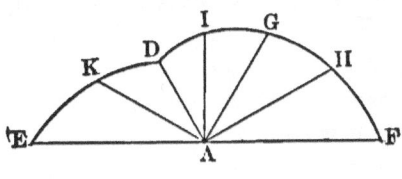

FIG. 31.

other half being like it. This wheel would work correctly with another like it, making a pair of 2-lobed wheels.

Again, by reducing the angles of Fig. 29 to a third their value, the third of a 3-lobed wheel would be obtained, from which the full wheel could be obtained by uniting three copied sectors. Likewise a 4-lobed, 5-lobed, etc., wheel could be produced.

These wheels would work together in pairs of equal lobes, but not interchangeably.

2. *By assumed angles and limiting radii.* In Fig. 31 lay off $AE = AF$ as maximum radii, and AD as a minimum radius. Bisect the angle DAF by AG and find the length of AG as in Fig. 21, thus obtaining a point G in the arc DGF. Similarly bisect GAF by AH, find AH as by Fig. 21, giving H a point in the curve. The point I is found in the same way, and other points between, to the extent desired, when the curve FHD can be drawn in.

Likewise for sector DAE, giving the half of a 2-lobed wheel from which the whole wheel is readily formed either by copying the second half from Fig. 31 as rights and lefts with this half, making the wheel symmetrical with respect to EF as an axis of symmetry, or by swinging the copied half around on the paper 180° and joining it to this half EDF, giving a non-symmetrical wheel.

Pairs of 3-lobed, 4-lobed, etc., wheels may be thus produced, but they would not work interchangeably.

INTERCHANGEABLE MULTILOBED LOG-SPIRAL WHEELS.

1. *As derived from one spiral.*

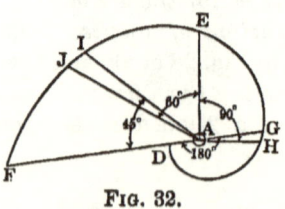

Fig. 32.

A 1-lobed wheel as in Fig. 28 will consist of two equal sectors of 180° each, as $DAGH$, Fig 32, or $FAGE$, etc., according to size.

A 2-lobed wheel will consist of four sectors of 90° each, as $EAIIG$, etc.

A 3-lobed wheel will require six sectors of 60° each, etc.

That these several wheels be interchangeable as derived from one spiral, Fig. 32, it is necessary that the sectoral arcs equal each other, while the sectoral angles differ.

Thus, in Fig. 33, the arc DCE must equal the arc FCG, while

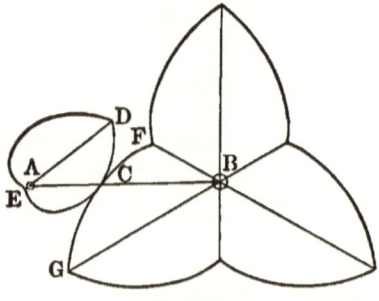

Fig. 33.

the angle $DAE = 180°$, and the angle $BFG = 60°$.

In Fig. 32 the 180° sector DHG gives half the wheel A, Fig. 33, while the sector EAJ, Fig. 32, gives a sixth of the wheel B, Fig. 33, these sectors in Fig. 32 being selected by trial measurements on a drawing-board so that the arc DHG = arc EIJ. Likewise the sector FAI, Fig. 32, would make the eighth of a 4-lobed wheel, and HAE the fourth of a 2-lobed wheel, any two of which series of wheels from the 1-lobed to the 4-lobed and upward will work together in pairs interchangeably.

The velocity-ratio is always equal $\frac{BC}{AC}$.

2. *As derived from two spirals of unequal obliquity.*

In Fig. 30 consider the wheel DEG as a 1-lobed wheel, to work with which a 2-lobed, 3-lobed, etc., wheel is desired.

Here it is only necessary to proceed with, say, the steepest spiral

first, just as was done in Fig. 32, obtaining sectors with arcs equal
that of the corresponding steepest sector of the one-lobed wheel,
and with angles a half, a third, etc., as great; resulting in the
steepest sectors of the 2-lobed, 3-lobed, etc., wheels; then likewise
with the flatter spiral, resulting in the flatter sectors, for the
2-lobed, 3-lobed, etc., wheels; when the unequal sectors for a lobe
of any one of the proposed wheels can be selected and the wheels
laid out.

Fig. 34 is an example of a 1-lobed wheel mating with a 3-lobed

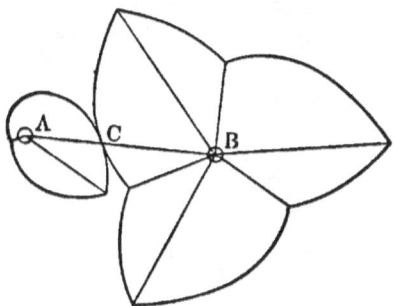

FIG. 34.

one laid out in this way. Any two of the lobed wheels in the
series above described will roll truly together.

3. *As determined from proportional sectors.*

In Fig. 35 let *ADEH* represent any given sector of a lobed

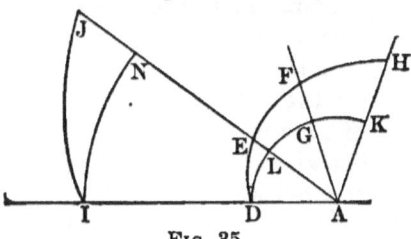

FIG. 35.

wheel. Required a sector *AIJ* of a wheel of three times as many
lobes that will work correctly with the first-named wheel. In this
case the sector *AIJ* must work correctly on the sector *ADEH*.

Now for three times as many lobes in the new wheel, new sectors of the new wheel which mate with the first must have the
sectoral angle one third that of the first wheel. That is, for the
sector *AIJ* to mate with sector *ADEH*, the angle *IAJ* must equal
a third of the angle *DAH*, and it is plain that the sector *AIJ* must

be similar to the sector ADE because of the equal obliquities. Also JN must equal HK, and IJ equal DEH. Therefore

$$EL : HK :: AD : AI :: AE : AJ,$$

a proportion which can be readily constructed graphically as in

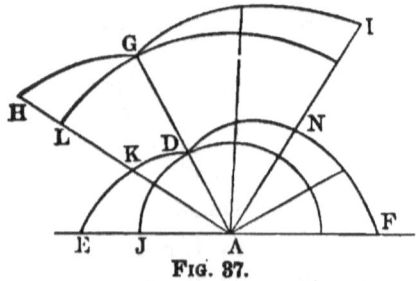

Fig. 36.

Fig. 36. With the mating sectors $ADEH$ and AIJ thus found, the wheels may be laid out.

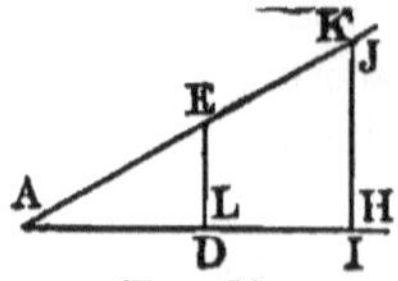

If in the first wheel the other sector for a lobe is not equal to the sector $ADEH$, but the two for a lobe form an aliquot part of the full circle of 360 degrees, then a similar construction will give the new mating sectors for the other parts of the lobes of the new mating wheel.

In this way the mating 4-lobed wheel is found for the given

Fig. 37.

2-lobed wheel A of Fig. 38 as shown in Fig. 37, where EDF is the half of the given wheel. The number of lobes being doubled, the angles will be halved, and the halves of the sectors of Fig. 37 being *similar* to the new sectors required, are simply expanded. Thus the half AKD is expanded to the full new sector AHG, making $HL = EJ$. Similarly the half of ADF, or ADN is expanded to the full new sector AGI, and the new 4-lobed wheel working with the former given 2-lobed wheel are both shown, mated, in Fig. 38.

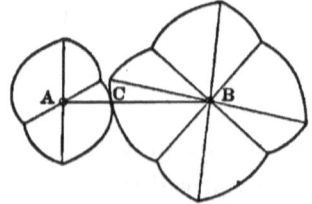

Fig. 38.

When one wheel is expanded to an indefinite number of lobes it becomes a straight notched bar, as in Fig. 39, of indefinite extent, the sectoral arcs of which are straight inclined lines where A may be the driver, C the point of contact, and the center B at an infinite distance away. The wheel B thus becomes a sliding bar DE. The driver A may have unequal sectors when the notches will be unsymmetrical like the teeth of a rip-saw.

The velocity-ratio is the same as that for Fig. 27.

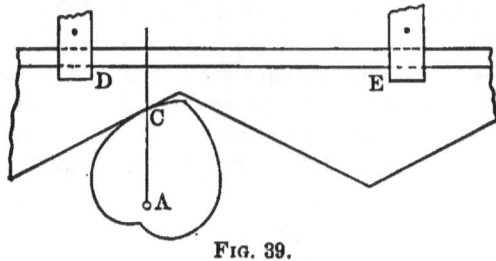

Fig. 39.

EASEMENT CURVES.

One characteristic of the above purely log-spiral wheels is the sharp abrupt intersection points, both salient and re-entrant, in the outlines where the sectors lie adjacent. These if desired may be *eased off* into curves of any form provided they will roll properly, as in Fig. 40.

Thus ADE is an easement sector mating with BHI, and AFG another mating with BJK. The easement sectors cannot be

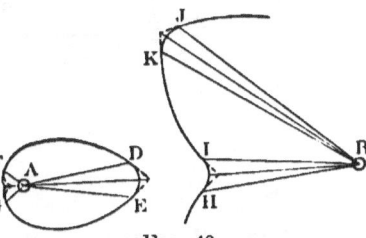

Fig. 40.

log-spirals, and they may be constructed by assuming one, as, for instance, DE, and then the mating curve HI must be found. How to thus construct mating curves is explained in a future chapter. To make such filleted outlines, DE and FG may be assumed, and the curves DF and GE drawn in as log-spiral sectors. Then as the position of the point B is not known exactly, an assumption may be made for it and tested by constructing a sector $HIJK$, which must be in proper relation to the full circle of 360 degrees.

Second. Equal and Similar Ellipses.

Fig. 41 presents a pair of equal and similar ellipses tangent to each other at C, with A, D, B, and E focal points. The ellipses are intentionally placed in contact so that the distance CH equals the distance CI, the points H and I being at the extremes of the major axes. If the ellipses be now rolled in mutual contact without slipping, in the direction such that H and I will approach, these points will meet when, at the same time, the major axes of the ellipses will coincide with one and the same straight line through AB.

30 PRINCIPLES OF MECHANISM.

It is readily seen that Fig. 41 is symmetrical with reference to the common tangent CO, and that $AC = EC$, $BC = DC$.

From a well-known property of the ellipse $AC + CD$ equals the major axis of the ellipse, equals FH, a constant for all points of contact C. But as CD always equals CB, we have $AC + CD = AC + CB = AB$, equal a constant, and equals the major axis of the ellipse. Likewise for BCE. Hence if we take the focal

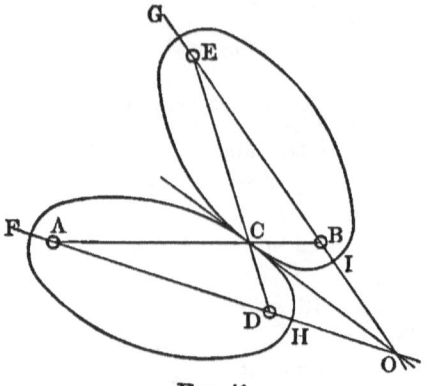

Fig. 41.

points A and B as axes of motion, the ellipses will roll truly in mutual contact as non-circular wheels.

For the same reason that AB is constant, DE is also constant, and hence pins placed at D and E may be connected by a rod or link with freedom of movement about A and B. It is interesting to note that the elliptical wheels transmit the same motion from A to B as would a pair of cranks AD and BE and a connecting rod DE.

The velocity-ratio for the elliptic wheels of Fig. 41 is 1 when the extremities of the minor axes are in contact at C, or when $AC = BC$; and least, or greatest in value, when the major axes touch at extremities on the line AB.

If A is driver and revolves uniformly, then the fastest movement of B occurs when H and I are in contact, and slowest when F and G are in contact; hence the ratio

$$\frac{\text{Fastest for } B}{\text{Slowest for } B} = \frac{\dfrac{\overline{AH}}{\overline{BI}}}{\dfrac{\overline{AF}}{\overline{GB}}} = \frac{AH}{BI} \cdot \frac{BG}{AF} = \frac{\overline{AH}^2}{\overline{BI}^2}.$$

ELLIPTIC WHEELS.

This ratio increases rapidly with eccentricity. For example, in a nail-driving device for a certain boot and shoe nailing machine (see Fig. 130) with elliptic pitch lines, $AH = 3.25$ ins. and $BI = 1$ inch, so that the fastest for B is 10.56 times the slowest; and the absolute velocity of A being usually about 300 revolutions per minute, the slowest rate of absolute movement for B is therefore $\frac{1}{3.25} \times 300 = 92.3$ revolutions per minute, while the fastest is $\frac{3.25}{1} \times 300 = 975$ revolutions per minute. The *ratio* of these figures is the same as by the formula given above, viz., 10.56. In this nail-driving device a crank is attached to B and by aid of a pitman or connecting link reciprocates a driving bar, the hastened motion of which bar resembled that of a driving bar accelerated by a spring, but possessed the great advantage of requiring no buffer to absorb the residual terminal shock, the crank and link always preventing the driving bar from going beyond its predetermined lowest position.*

Fig. 42 is an example from the Putnam Machine Co.'s exhibit at the Centennial of 1876, of elliptic gears used on a slotting machine to approximate the "quick return." The fastest motion of the driven wheel is nearly 4.5 times the slowest. The distance between centers was 14 inches.

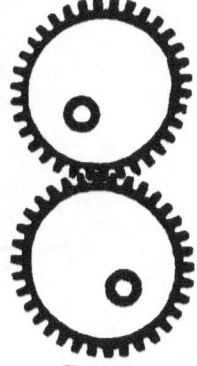

FIG. 42.

See also Fig. 133, for which the fastest for the driven wheel is 41 times the slowest.

---

* A general expression for the velocity-ratio is obtained from the equation of the ellipse, viz.:

$$AC = \frac{n^2}{m - a \cdot \cos \alpha},$$

$$BC = \frac{n^2}{m + a \cdot \cos \beta},$$

from which we find the

$$\text{velocity-ratio} = \frac{AC}{BC} = \frac{m + a \cdot \cos \beta}{m - a \cdot \cos \alpha}.$$

where α and β are the angles DAC and IBC; and where $2m$ is the major diameter and $2n$ the minor diameter, a being $\frac{1}{2}BE$.

PRINCIPLES OF MECHANISM.

The smoothness of outline of these elliptic wheels for rolling contact is favorable for their use in many cases where a variable velocity of the driven wheel is the main object, instead of some definite *law* of motion.

As the curves here considered are only for pitch lines of gear wheels, it is plain that angles in those lines are by preference avoided. Compare Figs 42 and 45 with Figs. 71, 73, and 89.

INTERCHANGEABLE MULTILOBED ELLIPTIC WHEELS.

The remarkable elliptic wheels of the Rev. H. Holditch are the only ones that may properly come under this classification. These are called elliptic, not because made up of parts of ellipses, but because of the peculiar relation of this series of wheel curves to a corresponding series of ellipses as shown in Fig. 43.

Draw the ellipse DPE with foci A and M and center O. Draw

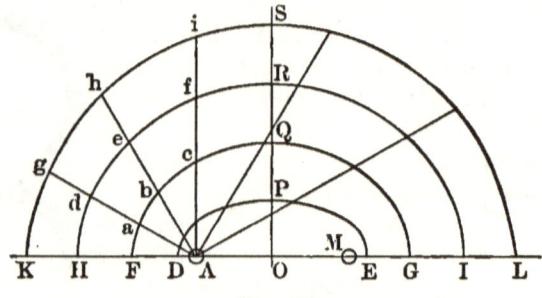

FIG. 43.

OS perpendicular to AM, cutting the ellipse at P. Then for complete multilobed wheels, lay off OP in repetition on OS, giving points Q, R, S, etc., so that $SR = RQ = QP = PO$. Then through these points draw confocal ellipses as shown, that is, with A and M as focal points. The points F, H, K, G, I, etc., may be found by making $OF = OG = AQ$, $OH = OI = AR$, etc., etc. Also we have $Aa + aM = AQ + QM$, etc. The full elliptic arc may be drawn by stretching a thread $AQMA$ around pins at A and M, and a pencil at Q. Then moving the pencil either way, kept tight against the thread, the ellipse is traced.

The ellipse DPE is the unilobe as at A in Fig. 44. The bilobe is obtained from FQG, Fig. 43, by using the radii AF, Aa, Ab, Ac, etc., laid off on lines radiating from B, Fig. 44, at half the angles FAa, aAb, bAc, respectively, Fig. 43, this giving a sector CBT, Fig. 44, which is the fourth part of the bilobe required.

Similarly the trilobe is obtained by laying off the radii AH, Ad, Ae, etc., on lines radiating from B', Fig. 44, at a third of the angles HAd, dAe, etc., Fig. 43, thus giving a sector $C''B'U$, which

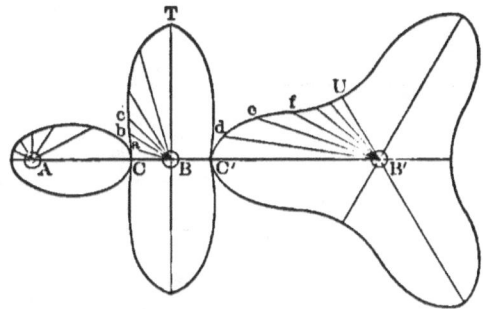

Fig. 44.

is the sixth part of the trilobe; from which the wheel can be constructed, as shown in Fig. 44.

In practice it is most convenient to make the angles equal to each other between lines radiating from A, Fig. 43; in which case the angles between the lines radiating from B, Fig. 44, are equal, and a half as great as in Fig. 43, and angles at B', Fig. 44, equal a third of those of Fig. 43, etc.

The unilobe A, Figs. 43 and 44, is an ellipse as stated. But the bilobe is not an ellipse, though somewhat resembling it. Again, the wheel B' is far from being an ellipse, and the reason these wheels are called elliptic is because of the use of ellipses in Fig. 43 in constructing them.

Fig. 45 is a photo-process copy of a pair of elliptic wheels like B and B' of Fig. 44, where the rolling curves serve as pitch lines for the teeth.

The demonstration of the Holditch wheels is not simple, and will not be given here, but the wheels are admitted at this point because of their direct dependence upon ellipses.

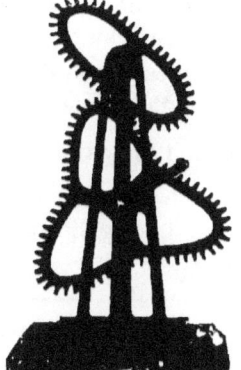

Fig. 45.

For the particular case that the number of lobes becomes infinite, the center of the wheel is removed to infinity, and what we see of it is shown in Fig. 46, where the curve becomes a sinusoid $LFHK$, in which $LH = \pi n$, and $GF = NK = \sqrt{m^2 - n^2} = AO$, if m and n be taken as semi-axes of the ellipse.

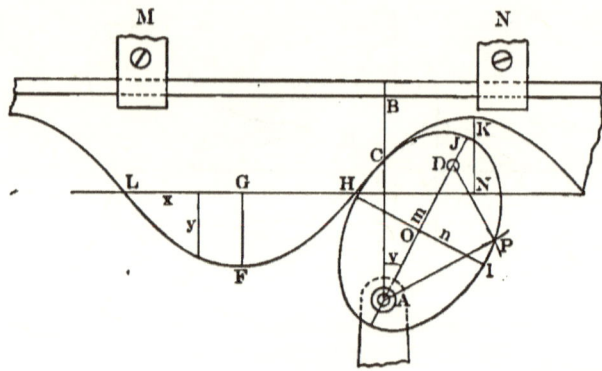

Fig. 46.

To construct the ellipse on a drawing-board: The half length and breadth m and n having been determined, and laid off on OJ and OI, make $IA = ID = m$. Then take any radius AP and describe an arc at P. Subtract this radius from $2m$, and with this difference as a radius, equal DP, describe an intersecting arc at P. The intersection will be a point in the ellipse. Any number of points may be thus found and the ellipse drawn in.

To construct the arc $FH = HK$, etc.: Take from a table of sines the values for every nine degrees up to ninety degrees, multiply these by $\sqrt{m^2 - n^2}$, and lay them off on ten equidistant perpendiculars between H and G and including the latter, through which points draw the curve FH, GH being $= \dfrac{\pi n}{2}$*.

The velocity-ratio is the same expression as that for Fig. 27, viz.: taking linear velocity of B as W, and angular-velocity of A as V, we have $\qquad AC \cdot V = W$,

---

* Where the semi-axes of the ellipse are m and n the equation of the curve of the sinusoid $LFHK$ is

$$\sin \frac{x}{n} = \frac{\pm y}{\sqrt{m^2 - n^2}},$$

found by integrating the polar equation of the ellipse for $dx = rdv$ and remembering that $r = m \pm y$.

or
$$\frac{W}{V} = Ac,$$

making the rate of linear movement of B along the guides M and N the same as that of the end of a radius AC for the instant as revolving about the center A.

Third. A Pair of Equal and Similar Parabolas.

A parabola may be regarded as an ellipse so extended that one focus is at an infinite distance away. Thus in Fig. 41 suppose the focus B and D to be infinitely removed. Then F and G are vertices of parabolas, with A one axis of motion while the other, B, is at an infinite distance. In this case the motion of BG is simply to slide, as for B in Fig. 46. These rolling parabolas are shown in

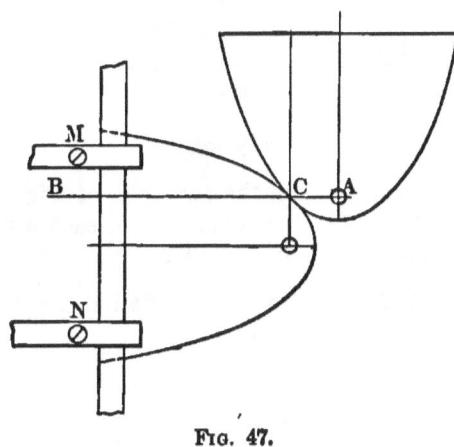

FIG. 47.

Fig. 47, where A is the fixed axis of revolution for wheel A, while B slides in guides M and N.

To prove that the parabolas will thus roll truly, we make use of Fig. 48 of two equal parabolas with their vertices in contact at C, and with A and O focal points, A serving as an axis for A, and with B sliding in guides M and N. A property of the parabola is that the radial line AD makes the same angle as does DF with the tangent at D, or OF with the tangent at F. Also a well-known property is that the line AD equals the line DG equals FH, where OG and AH are perpendicular to AO. Now with A, M, and N fixed, suppose A to turn until D falls at K, at the same time sliding OB along. Then F will be brought exactly to K, with the latter a point of rolling contact on the line of centers AOK, as shown by the dotted lines, because $AK = FH = GD$, and the angles between the tangents to the parabolas at K and the line AK are equal to the angle between AD and the tangent at D. The same is true for all

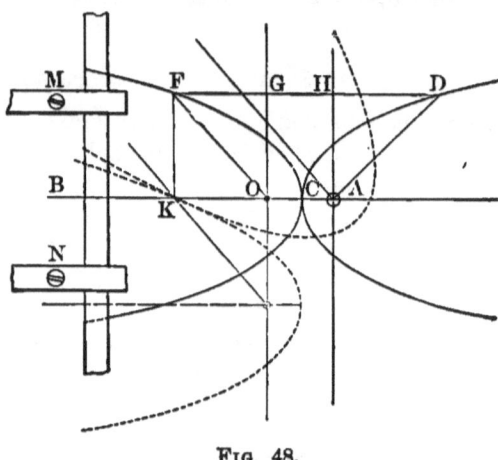

Fig. 48.

the points of contact between the parabolas of Fig. 48, and hence the parabolas of Fig. 47 will roll truly upon each other, A turning about its pin while B slides in M and N.

Fourth. **A Pair of Equal and Similar Hyperbolas.**

In Fig. 49 let LCM and ICJ represent a pair of equal and simi-

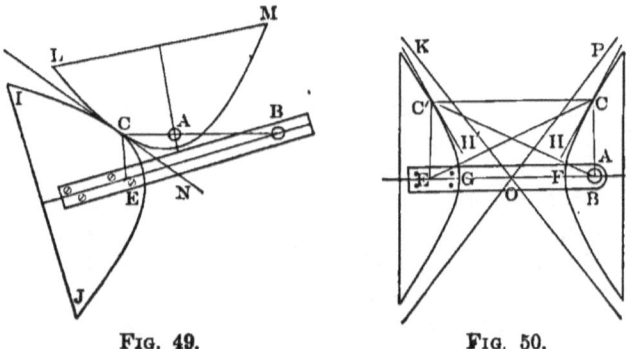

Fig. 49.   Fig. 50.

lar hyperbolic sections for which A and E are focal points; the focus A being taken for the axis about which LCM rotates, and B the axis about which ICJ turns, the latter being mounted by a bar

BE, made fast to it, and centered by a pivot at B. A and B are then centers of motion, and BAC the line of centers, which must of course constantly pass through the point of contact C.

To prove that these hyperbolas will roll truly, we refer to Fig. 50 of the same hyperbolas presented in correct relation with their asymptotes OP and OK, and focal points A and E. A well-known property of these curves is that the difference in length of any pair of lines AC and EC from any point C is the same for all points C on the curves, and is equal to the distance FG between the vertices and constant. Another property is that the tangent CH at C bisects the angle between the radial lines AC and EC.

Now if we suppose the hyperbola ACF to be lifted from the paper as drawn, and be swung around till HC coincides with $H'C'$, and the point C with the point C''; then the hyperbolas will be in common tangency at C'', and the distance from the new position of A over to B will just equal $BC'' - AC = FG, = AB$ of Fig. 49; and the two figures will otherwise agree; presenting conditions consistent with true rolling contact.

The velocity-ratio is

$$\frac{V}{v} = \frac{BC}{AC}.$$

In this case the point of contact is outside the space AB, and at any instant the wheels revolve in the same direction, while in former cases the contact is between the axes and the motions contrary.

Fifth. Transformed Wheels.

Let Fig. 51 represent any pair of correct rolling non-circular wheels; in this case sectoral.

Divide the two wheels into *mating elementary sectors*, for which a couple of mating short arcs may be referred to as a *pair* of arcs; a mating couplet of radii as a *pair* of radii; and a mating set of angles as a *pair* of angles.

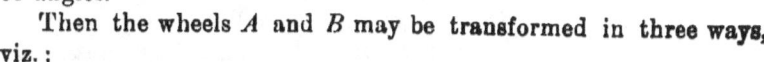

Fig. 51.

Then the wheels A and B may be transformed in three ways, viz.:

1st. By multiplying each of a pair of angles, or of arcs, by a constant. (Holditch.)

2d. *By adding a constant length to all radii of one wheel, the arcs remaining constant.*

3d. *By adding to one and subtracting from the other of a pair of radii a given quantity; arcs remaining unchanged.*

In these changes the elementary, or small, sectors are supposed to have angles of from three to ten degrees, according to desired accuracy of result, the precision of mathematical analysis not being expected, though a sufficiently close approximation to it is obtained to answer for most practical problems arising under this proposition.

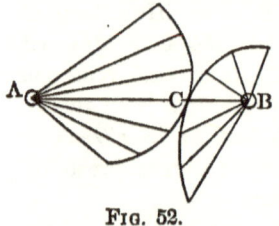

Fig. 52.

The second mode of transformation, applied to the wheel B of Fig. 51, by cutting off BB' from all radii, or from B to the circle DE, that is, adding negatively the length BB' to all radii of B, and bringing the cut-off ends of the remaining portions of radii all to the point B', will give the pair of wheels shown in Fig. 52 which will be found to answer the conditions of true rolling contact.

The first mode of transformation applied to the wheels of Fig. 52, by multiplying all the angles by two for a simple case, changes those wheels to the ones shown in Fig. 53. In this multiplying it is not necessary to use the same multiplier for all pairs of angles; indeed a different multiplier might be used for each pair of angles.

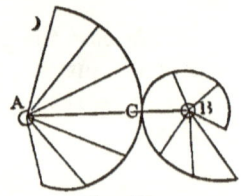

Fig. 53.

The third mode of transformation applied to Fig. 52, results in Fig. 54, where Aa is shortened and Bc lengthened the same amount; Ab shortened and Bd lengthened the same amount; Ae lengthened and Bg shortened the same amount; and finally, Af shortened and Bh lengthened the same amount, these changes of the radii being made without any alteration of the lengths of the elementary arcs Ca, ab, Cc, cd, Ce, Cg, etc. It will be understood that the positions of the radii will be changed, that is, that the angle CAa will be varied as Aa is shortened, while the length Ca is preserved; and likewise for the other angles and arcs.

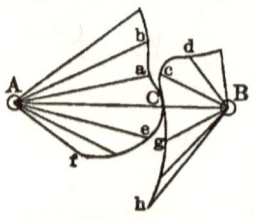

Fig. 54.

These three modes of transformation are sufficient to change any pair of correct rolling wheels into any other form. Thus, starting with the sectoral wheels of Fig. 51, suppose it were sought to derive a pair of *complete* non-circular wheels.

Then, first, make the limiting radii of A equal, the one to the other, also B, by the third mode; then by multiplying pairs of angles by a given quantity, one pair after another, as by the first mode, we may make the total angle of one of the sectors 360 degrees, giving a complete wheel, while the other will most likely not be complete. To make the latter complete also, apply the second mode of transformation until that wheel closes with 360 degrees.

That the wheels thus rendered complete shall have the necessary shape as to eccentricity, law of varying velocity-ratio, etc., the wheels may be examined before entirely closed, and so treated by the modes of transformation as to bring about the desired law of velocity-ratio of the completed wheels.

Extended changes of wheels in this way in general become tedious, and other processes on the drawing-board, hereafter explained, will be preferred. The above modes of transformation will, however, be found most useful in making minor changes, such as rounding off a sharp prominence or depression, as in Fig. 40; and in certain special cases.

INTERCHANGEABLE MULTILOBES BY TRANSFORMATION.

The transformation of a pair of correct working wheels into another pair, wider or narrower, by the first mode, is a comparatively simple operation.

Thus, in Fig. 55, we have a pair of *Transformed Elliptic Wheels*, mating half-ellipses A and B, in dotted lines, in which the angles are reduced by one half

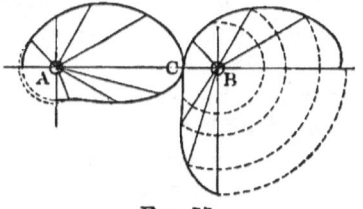

Fig. 55.

without changing the radii by which the 180-degree sectors are reduced to 90-degree sectors, or the one-fourth part of a pair of 2-lobel wheels, as explained in Willis, page 65.

Other portions of rolling ellipses may be transformed into lobed wheels of intersecting characteristics, as explained in MacCord's *Mechanical Movements*, page 54.

As a second example, take the equal elliptic wheels A and B, Fig. 56, and transform A into a wheel of two lobes as at A', by adding a constant quantity to all the radii of wheel A, the elementary arcs remaining unchanged, as per the second mode of transformation. We will transform the half of A, the other half being like it except as to "rights and lefts."

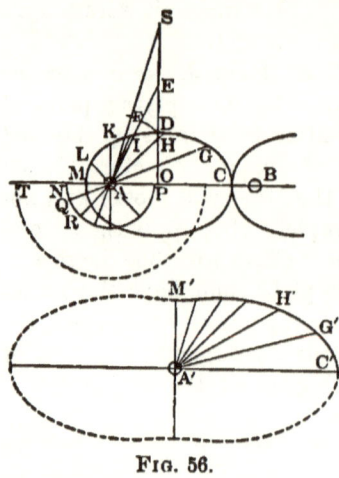

FIG. 56.

The amount to add to all the radii of A to obtain a 90-degree resulting sectoral wheel $A'C''M'$ to give exactly the fourth part of the wheel A' is not readily determined, several trial values being in some cases necessary. In the present case, however, we may refer to the Holditch principle of Fig. 43, from which we find that by subtracting the distance, Fig. 56, AD from the distance AE we obtain EF for the amount to add to all the radii CA, GA, HA, etc. To apply this conveniently, draw the circle NRP to center A with radius EF.

Then to construct $A'C''M'$ take $A'C' = CN$, thus adding AN to AC. Also make $A'G' = GQ$, and $C'G' = CG$, thus determining the point G'. For the point H' make $A'H' = HR$, and $G'H' = GH$, thus determining the point H'. Likewise proceed to M', making $A'M' = PM$. Trace the curve through the points $C''G'H'\ldots M'$, completing the 90-degree sector $A'C''M'$, with which the complete 2-lobed wheel A' may be laid off. This wheel A' will work correctly with the wheel B as a pair of 2-lobed and 1-lobed wheels, as A and B of Fig 44.

The quarter wheel $A'C''M'$ is identical with the Holditch wheel found by the principles of Figs. 43 and 44, though the points G', H', etc., will not coincide with the Holditch points even for the same number of points, and the same angle CAG, GAH, etc.

To make a 3-lobed elliptic wheel by this mode of transformation make $ES = ED$, and subtract AD from AS, which difference use as the radius AT of a circle about A, from which measure the radii for the 3-lobed sector similarly as in the 2-lobed wheel. Combine these radii with the same arcs CG, GH, etc., for points in the 60-degree sector for the sixth part of the 3-lobed wheel.

Fig. 28 could be used as basis of a series of multilobed wheels which would work interchangeably, spiral arcs and differences of limiting radii remaining unchanged.

TRANSFORMED PARABOLIC WHEELS.

As an example of a decided change in the appearance of a wheel take those shown in Fig. 48, and move the center of motion from A to A', Fig. 57, thus adding the distance AA' to all the radii, the elementary arcs being preserved in length.

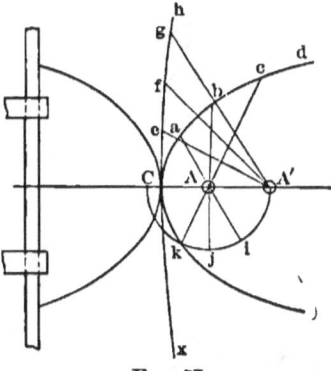

Fig. 57

Divide the arc $Cabd$ into equal parts and draw the radial lines aAi, bAj, etc.

Then the radius of the point C of the transformed wheel will be $A'C$, and of the point e it is $A'e$, equal to aAi, where arc $Ce = Ca$; also radius $A'f = bAj$, and arc $ef = ab$, etc.

The arc hCx of the transformed wheel, when $CA' = \frac{1}{3}CA$, will be nearly a straight line, but not exactly; because to be straight requires the contour of B to be a logarithmic curve resembling the parabola.*

WHEELS OF COMBINED SECTORS.

An interesting series of wheels may be obtained by combining sectors of the above various forms, as, for example, of the ellipse and log-spiral.

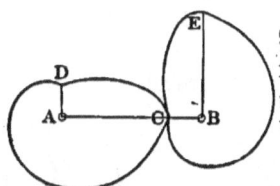

Fig. 58.

In Fig. 58 we have for wheel A a 90-degree elliptic sector ACD like the fourth part of B, Fig. 44, while the remaining 270-degree sector is a log-spiral. Wheel B is a copy of A.

In Fig. 59 we have a 2-lobed wheel A, each lobe of which is composed of a 90-degree elliptic sector like BCT, Fig. 44, and a 90-degree log-spiral sector; working with a 3-lobed wheel B,

---

* The equation of the logarithmic curve of B, making the mating curve hCx a straight line, is $y = a \log \dfrac{\sqrt{x^2 + 2ax} + x + a}{a}$, where x and y are co-ordinates of the curve BC, and $a =$ the distance $A'C$. The polar equation of the outline of the wheel A' is $r = a \sec v'$, v being angle $CA'e$, etc.

each lobe of which has approximately a 60-degree elliptic sector like $B'C'U$, Fig. 44, and approximately a 60-degree log-spiral sector. If wheel A is first made with 90-degree sectors wheel B is best laid out by trial values of the sectoral angles, since the angles

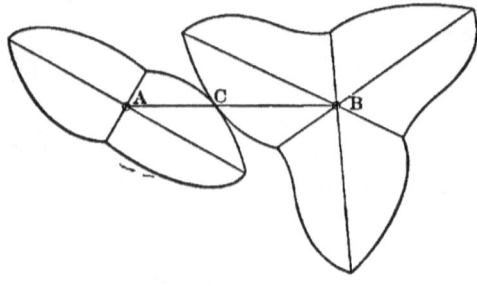

Fig. 59.

will not be exactly 60 degrees in B. When a lobe of 120 degrees is found by trial for B, the remaining two lobes can be copied from it.

In Fig. 60 DFC and CGE are a pair of rolling half-ellipses with axes of rotation at A and B. The other half of each wheel consists of log-spiral sectors ADH and AJC, separated by a circular sector AHJ, laid out by first assuming the circular sectoral angle $CAI = HAJ$, then drawing in the log-spiral DHI, according to Figs. 20 and 21, then finding the middle point H, and transferring the sector AHI to AJC, when the circle arc HJ completes the drawing of the half-wheel A. The mating half-wheel of B is a copy of $ADHJC$.

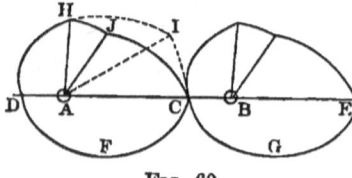

Fig. 60.

In Fig. 60 the wheels, though composed of several sectors, are still similar, while in Fig. 61 we find several sectors combined in wheels that are quite dissimilar. Here CD and CG are a pair of elliptic sectors, DE, EF, and GH, HI log-spiral sectors, and CF and CI elliptic sectors.

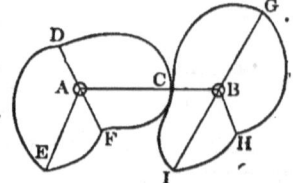

Fig. 61.

The above examples will serve to illustrate the fact that a great variety of wheels may be made from the four curves named as the foundation.

There are many cases, however, where none of the above wheels will answer, as, for instance, where a particular and definite *law* of angular velocity foreign to the above curves must be closely followed during a part of, or even throughout, the entire revolution, while the above wheels have laws of motion of their own which will rarely fit the special case of a predetermined law.

The above wheels are fairly well adapted to the simple requirement of a specified maximum or minimum velocity, or both, as occurring within a revolution, and at certain points therein.

CHAPTER III.

NON-CIRCULAR WHEELS IN GENERAL.

In modern practical mechanism there exists a demand for non-circular gearing where the *law of motion* of the driven wheel is fixed by considerations foreign to log-spirals, ellipses, etc.; which law the required wheels must follow, sometimes approximately, and sometimes *exactly*, in some cases for a part of, and in other cases throughout the entire revolution of the driven wheel.

Again, there may be demanded *similar wheels*, differing from log-spirals and ellipses, and which can be readily drawn, and readily changed and redrawn to suit the designer's fancy. Hence there arise three cases, viz.:

I. *One wheel given, or assumed, to find its mate ;*
II. *Laws of motion given to find the wheels ;* and
III. *Similar wheels.*

In either case the wheels may be complete or incomplete. Treating these separately, we have:

I. GIVEN ONE WHEEL TO FIND ITS MATE.

In Fig. 62, let A represent the assumed wheel so chosen that when the wheel B is found its circumstances of motion will approximate those desired, more or less closely according to judgment in assuming A. When B is completed and tried, and found not to move as desired, A may be modified according to dictates of the first trial, and B found again.

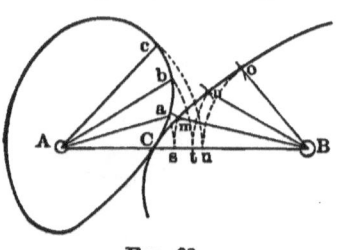

FIG. 62.

In drawing the wheels the outline $Cabc$, etc., may be traced, and the center of motion, A, arbitrarily assumed.

Then assume a point B for the center of motion for the wheel $Cmno$, and draw the *line of centers* AB. C is one point in the outline of B. Then describe an arc am, produced, with C as the center. Also subtract Aa from the line of centers AB, and with the remainder, equal Bm, laid off from B, find the point m, as the second point in the wheel B. In other words, with Aa as a radius, and the center A, describe the arc as. With B as a centre and Bs as radius, describe the arc sm, when the intersection m is the second point in B. Next, with ab as radius and m as centre, describe the indefinite arc at n; and with Ab as radius, describe the arc bt, and with Bt as radius, describe the arc tn, to find the intersection n, as a third point in the wheel B.

Thus we have $Cm = Ca$, $mn = ab$, $no = bc$, etc.; and $Bm = AB - Aa$, $Bn = AB - Ab$, etc.

Proceed thus for the entire circumference of A, resulting in corresponding or mating points for B. A curve traced through these points gives the first trial outline for the wheel B.

Now if, when the center point B is assumed, it makes the line of centers AB too great, the wheel B will have a gap in it; and conversely if AB be taken too small there will be an overlap.

To make B complete, that is, to just close up without a gap or overlap, a new point for B must be assumed, and the whole work of finding the points mno, etc., repeated, one or several times, until sufficient closeness is obtained, and the outline of B traced, thus completing the pitch lines, or wheels, A and B.

The distance to move B will be not far from one-sixth of the arc of the gap, or overlap.

The above tentative process, as unscientific as it may appear, is nevertheless the only way for the case of one wheel given to find its mate.

As to accuracy, to make the spaces $Cabc$, etc., excessively great introduces an error due to the difference between the *chord* and *arc;* while to make them excessively small, the errors of graphical work accumulate. Probably 10-degree angles around A will be found about right.

To increase the length of the steps Ca, ab, etc., without jeopardizing accuracy, Rankine makes the ingenious application of the following

Graphic Rules for Equivalence of Lines and Arcs.

First. *To find a straight line equal to a circle arc.*

To find the tangent line $DF =$ the arc DE, Fig. 63, take $DO = \frac{1}{2}DE$, and with O as a center, describe the circle arc EF, cutting DF in F. Then $DF =$ arc DE within less than a tenth of one per cent, up to an angle of 30 degrees, and increases as the fourth power of the angle.

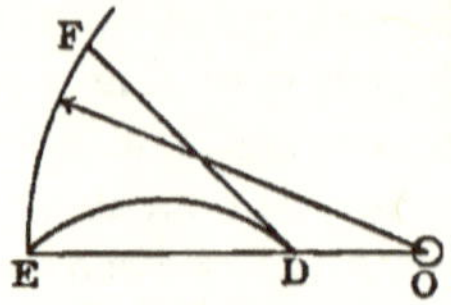

Fig. 63.

Second. *To find a circle arc equal to a straight line.*

To find the arc $DE =$ line DF, Fig. 64, take $DO = \frac{1}{2}DF$, and draw the circle arc FE from O as a center. Then arc $DE =$ line DF within same limits as first rule above.

The arc DE must be tangent to the line DF at D. Hence the center of the arc DE is on DG drawn perpendicular to DF at D. The same rule and construction apply for all arcs starting from D, with centers on line DG, as for the dotted arcs shown.

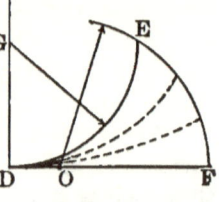

Fig. 64.

The above percentage of error is true for circle arcs only; other curves give rise to greater errors.

To Draw a Tangent to any Curve.

To draw a tangent line to the curve DOE at the point O.

First draw any curve FG, cutting the curve DOE, and also a series of straight lines as DI, IIe, ac, etc. Then lay off the length DF from O to I, noting the point I; Oe from G to H, noting the point H; ao from b to c, noting the point c, etc. Then trace a curve through the points $HcdI$, etc., as shown, noting its intersection T with the curve FG. Then a straight line through the points T and O will be tangent to the curve DOE at O.

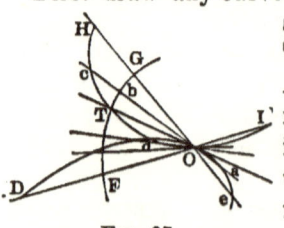

Fig. 65.

These rules will be found useful in the graphic operations of problems in Mechanism.

Application of Rule Second to the problem above, of *one wheel given to find its mate*. Here $ACDE$, Fig. 66, is a portion of wheel A, and B the proposed center of wheel B, with AB the line of centers.

Draw Cd tangent to the wheel A at the point C. With a as a center, taking $Ca = \frac{1}{4}Cd$, draw the circle arc DdF extended. Draw the circle Dm from A, and the circle mF from B, giving the intersection F of arc mF with arc DdF. Then F is a point in the periphery of B, since arc $CD =$ line $Cd =$ arc CF, of B, according to rule 2d above.

To find another point in B, draw the tangent De at D, and the radius AD extended, noting the angle gh. Then draw the radius BF, and the line Ff at the angle with FB of $ij = gh$. Take the points c and b at the $\frac{1}{4}$ distance De from D and from F. Then

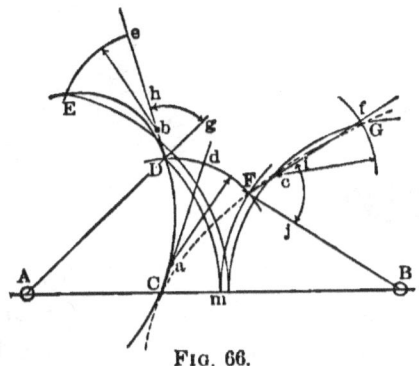

FIG. 66.

with be as a radius and center b, draw the arc eE; also from center c the arc fG extended. Then, with a radius $AB - AE$ and center at B, cut the arc fG at G. This latter point is another point in the contour of wheel B, since arc $DE =$ line $De =$ line $Ff =$ arc FG of B.

Thus proceed for the whole circumference of A; and if the circumference of B closes, as by center B having been rightly chosen in distance from A, the points C, F, G, etc., will be correct points in the contour of B. But if wheel B does not close, assume another position for center point B and repeat the work for a new set of points C, F, G, etc. Finally when the wheel B closes, the points C, F, G, etc., will be correct points in the arc of B, though they will be too few to admit of drawing in the contour of B.

A sufficient number of points intermediate may then be found as by the method of Fig. 62, thus completing the wheel B.

The advantage of the wide steps of Fig. 66 is to permit of finding the correct position of the center B with comparatively little work.

In Fig. 67 is a photo-process copy of a pair of wheels in metal, the pitch line of the lower one of which was first traced by pencil free-hand, and the center point assumed, when the upper or mating wheel was found by the process above described. The wheels are about 6 inches between centers.

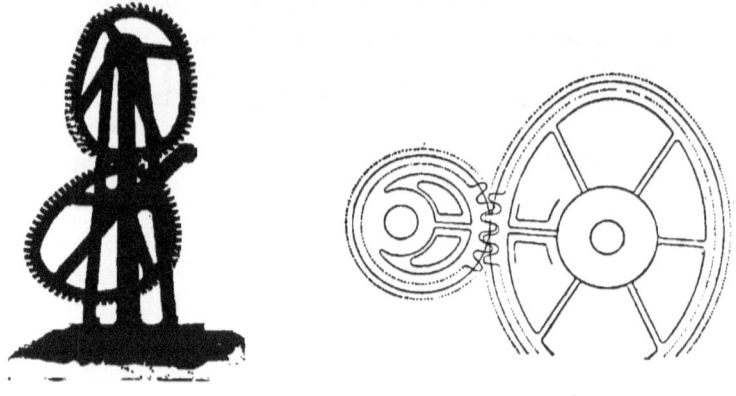

FIG. 67.   FIG. 68.

Fig. 68 is a drawing copied from a pair of wheels used on a paper-cutting machine from England and exhibited at the World's Fair of '93. The wheels were in cast iron, about 16 inches between centers, in which the smaller is simply an assumed circle, with its axis chosen arbitrarily in eccentricity, while the larger wheel is 2-lobed.

II. LAWS OF MOTION GIVEN TO FIND THE WHEELS.

The centers of a pair of wheels chosen to illustrate this case are taken at A and B, Fig. 69, with AB the line of centers, and the heavy outlines $CDFH \ldots X$, and $CEGI \ldots Y$, the contour lines of the completed half-wheels, or at least as much as lies above the extended line of centers.

The wheel A in Fig. 69 is supposed to revolve about its center of motion with *velocity constant*, so that the "law of motion" for A is very simple, viz., *motion uniform*. A semi-circle $aceg$, etc., is drawn within the half-periphery of A and divided into six equal angles or sectors each of 30 degrees. For convenience suppose the wheel A to make a half-revolution in six seconds, or any one of the six sectors in one second. Then the law of motion of A is represented graphically by the sectors into which it is divided, these several sectors passing over the line of centers in equal times, say one second for each. These angles and sectors may be called velocity angles or sectors, or sometimes auxiliary angles or sectors.

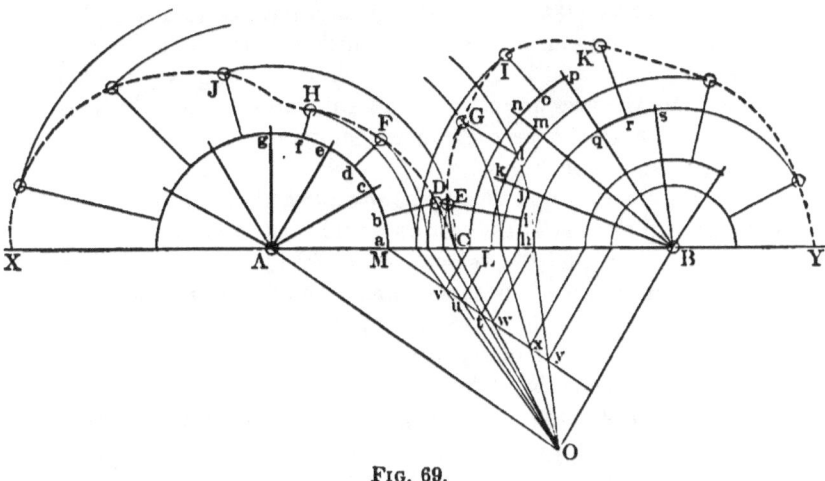

FIG. 69.

If for B the sectors were equal also, its motion would be uniform, and the resulting wheels would be circular and without special interest. But let the sectors of B be varied in angle as shown, any one sector passing the line of centers per second. Then the law of varied motion of B may be represented graphically by

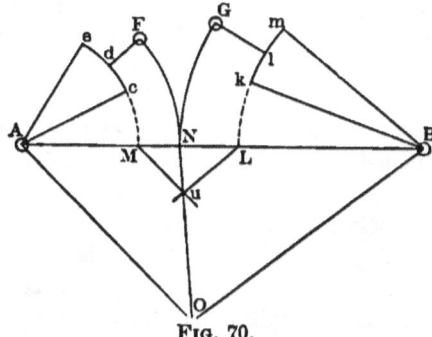

FIG. 70.

the varied sectors, as shown. Hence the laws of motion of the wheels are represented by the angular spaces laid off around the centers A and B. Though A, here, has equal angles and uniform motion, it may readily have a varied motion and a correspondingly varied series of sectors, as for the case that the driver A has a varied motion. But in practice the driver will usually have uniform motion.

Next, suppose all the sectoral angles to have sectoral arcs drawn in, these arcs being all of the same length, so that the narrower the sector the greater will be its length. Thus the angles for A being all the same, the arcs are all the same distance from A in a circle arc. But for B the arcs arrange themselves at varied distances from B, as shown in Fig. 69.

The sectors of A and B may be regarded as in pairs, so that if ac should roll against hj the sectors would turn through their angles in the same time, and the velocity-ratio would be the inverse ratio of the sectoral radii. The same being true of the next pair of sectors rolling upon each other, it follows that the velocity-ratio would change from the one pair of sectors to the next, in case the radii differ, as in fact they do.

Thus for A moving uniformly, B would have a varied motion according to its sectoral angles, or according to the stated law of motion, since the angles were laid out according to that law.

Hence if these wheels, composed of stepped sectors, could be kept revolving with these sectoral arcs in rolling contact one after another in order, the law of motion, in general, would be correct. But this would be thoroughly impracticable, since it would require the distance between the centers A and B to be continually varying. This is obviated by changing the radial lengths of the sectors so that their ratio in pairs will be the same as before, and their sum equal the line of centers AB, while at the same time the modified sectoral arcs are merged into each other in continuous lines, as shown in the lines $CDFH\ldots X$ and $CEGI\ldots Y$.

Fig. 70 will serve best to show how this may be done, where A and B are the centers, cAe a sector for wheel A, and kBm a sector for wheel B, where are $cde =$ arc klm, these sectors with lettering and construction lines being taken from Fig. 69. Now to find points F and G in the final perimetric arcs of wheels A and B, such that the velocity-ratio for F and G in driving contact shall be the same as that for the arc cde rolling on the arc klm; it is only necessary that

$$\frac{\text{length } AF}{\text{length } BG} = \frac{Ad}{Bl}.$$

By graphic process this is most readily done by drawing lines from A and B to meet in some point O, and parallels thereto from

M and L meeting in u, where cM is an arc struck from A, and kL an arc struck from B, and a line OuN, when

$$AM = Ad : AN :: BL = Bl : BN.$$

Now with radius AN find the intersection F by drawing the arc NF; likewise with the radius BN find the intersection G; dF being drawn from the center point A and the middle of ce, and Gl likewise drawn with respect km and B.

Fig. 70 is taken out of Fig. 69, with like lettering; and by referring to Fig. 69, we find that other points, as DHJ, etc., and EIK, etc., were obtained in the same way. Through these points, when all are thus determined, the contours of the completed half-wheels A and B may be drawn in with smooth outlines.

In practice the angles of the sectors should not exceed 10 to 15 degrees for suitable accuarcy of results.

In the figure we have drawn only half of each wheel A and B, but of course the remaining halves may be drawn in the same way.

In practice these wheels have been made where the specific law of motion of B was carried in some cases through 180 degrees, and in others through the whole 360 degrees. At the Centennial of 1876 there was exhibited, by B. D. Whitney of Winchendon, Mass., a certain barrel-stave sawing machine where the carriage carrying the blank block for staves was fed against the "tub-saw" at a uniform rate of advance and with a quick return; the movement of the carriage being made by a crank and pitman connection driven by non-circular gears of Fig. 71, $15\frac{1}{2}$ inches in diameter. On examining the marks of the saw on sawn staves it was found that the forward feed was wonderfully close to a uniform rate of advance, proving that the wheels must have been carefully worked out as to laws of motion, even allowing for a short pitman.

In this case the wheel A made something like a three-fourths turn, while B made a half-turn with uniform motion of stave block.

Fig. 71 is an accurate copy, the wheels serving as type, which, with printer's ink and paper, gave the copy that is photographed for the figure.

The divergence of the teeth near the salient angles is noticeable, and suggests *blocking* of the gears, and yet they actually operated in the most satisfactory manner.

Also see Fig. 120 for another example of careful laying out.

In another example of these wheels the pitman was in effect of infinite length, giving uniform motion to a carriage forward, with

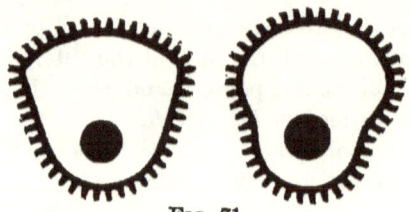

Fig. 71.

a quick return, the driver revolving uniformly for about a three-fourths turn, to the half turn of *B*, as required in a boot and shoe screw nailing machine. See Figs. 134, 135, and 136.

As a case of quick return with short pitman with a little more latitude for return, and for starting and stopping at the ends of the uniform motion, see Fig. 72.

Fig. 72.

At the Centennial the problem was made known of a motion for winding yarn on a conic bobbin, such that the yarn would be taken off a reel uniformly when winding upon the cone from a base of four inches to a tip of one inch, back and fourth on the cone in regular screw pitch. This requires the cone to revolve faster when winding yarn on its tips than when winding on its base, and *vice versa*.

This problem has since been worked out in unilobed non-circular gears, such that while the driver A revolves uniformly, the follower B will have an acceleration for a half-revolution during the windings down the cone, and then a retardation for a half-revolution of B during the winding up the cone. Thus one revolution of B takes place while winding down and back the cone. Now if the cone is to have several convolutions of yarn, say ten, in winding down, and then ten in winding back, the cone must make twenty revolutions to one of B. This relation can be obtained by interposing circular gearing between the cone and B, with a ratio of increase of speed of twenty to one.

Fig. 73.

These non-circular gears, as unilobed wheels in metal, are shown in Figs. 73 by photo-process copy.

In Fig. 89 these same wheels are constructed as bevel non-circular wheels.

In Figs. 76 to 78 the question of determination of the pitch lines of the non-circular wheels for this bobbin-winding problem is fully solved as a particular example in illustration of the application of the principles of Figs. 69 and 70; the example answering to the case of *law of motion definite and unalterable for the entire revolution.*

Even this is simpler than the most general possible case where the law of motion for the second half of the revolution of the driven wheel is different from that of the first half, as it may be; for which case the resulting wheels are non-symmetrical, and may resemble the wheels of Fig. 67, for which wheels, of course, assuming the driving wheel in uniform motion, the driven wheel follows some law of motion, simple or complex, which law, however complex, may be made out and expressed graphically, where, instead of using velocity sectors as in Figs. 74 and 78, passing time may represent the abscissas, and where the ordinates represent velocities of the driven wheel for the corresponding moments of time or corresponding points of revolution of the driver.

As an example of slightly unsymmetrical wheels, suppose the tool of Fig. 75 were required to return at a uniform rate of motion, as well as to advance according to that law. That return may be in the same time as the advance, for which case, if B is in the prolongation of the line SU, the wheels would be symmetrical.

But they would not be symmetrical when taking B above SU as shown; and still more unsymmetrical if the return of the tool, still uniform in velocity, were to be in less or more time than the advance. For this, according to the description of Figs. 74 and 75, the scheme of velocity angles of the lower half of B is to be made in the same way as the upper half shown. Then A is to be divided into two parts, apportioned according to the assignments of time for the advance and the return of the tool T, those parts to be divided into velocity sectors corresponding in numbers with those of the mating parts of B.

SOLUTIONS OF SOME PRACTICAL PROBLEMS.

FOR LAWS OF MOVEMENT GIVEN TO FIND THE WHEELS.

As this subject is evidently one of considerable importance in *practical mechanism*, the solution of several problems is here given to illustrate the application of the above principles.

1st. The Shaping-Machine Problem.—This is called the *shaping-machine problem* because on the machinist's shaping machine the cutting tool is desired to have a *uniform motion* forward for the cutting stroke, and a *quick return* stroke where the tool returns idle and should do so quickly to save time.

For this we have given a shaft, A, revolving uniformly, gearing with a second shaft, B, revolving at such variable rate of rotation as to secure the uniform forward and quick back stroke as described; also a carriage for the tool connected to the driven wheel B by pitman and crank.

In Fig. 74, T is the tool to move along the cut QR, held on the slide ME and guided by NO. The slide EM, carrying the tool, is connected to the crank BD by the pitman DE, E being the "crosshead" pin moving in the line or path SU, which pin moves with the tool.

Divide the path SU into equal parts to represent uniform motion, the pin passing over these parts forward in equal times. The pitman $DE = FS$, $= HU$, $= GV$, etc., as shown. Hence the crank pin D must move through the arc spaces FG, GD, DI, IJ, JK, KH, in equal times, from which we have the angles FBG, GBD, DBI, etc., to represent the law of motion of the driven wheel B for the forward stroke. The back stroke being made quick, simply, we may assume the angles HBL, DBP, and PBF, which for the quick return should be larger than those for the forward stroke.

Thus, by aid of Fig. 74, we obtain the angles to be passed over

SOLUTIONS OF SOME PRACTICAL PROBLEMS. 55

by the wheel B for its entire revolution. Taking WX for the line of centers through A and B, we may transfer the velocity angles to Fig. 75 with like letters.

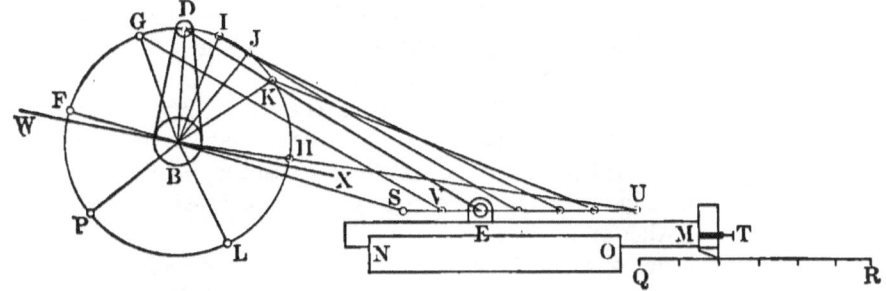

FIG. 74.

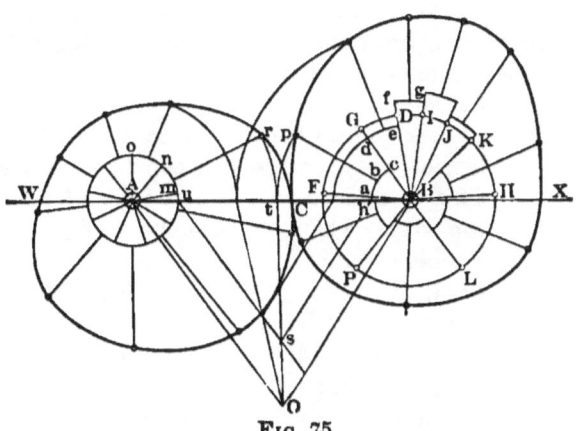

FIG. 75.

In Fig. 75, as was done in Fig. 69, draw equal arcs in the angles about B, as $ac = de = fg$, etc., and extend radii from their center points as shown, nine in all. Taking A to revolve with uniform velocity, there will be nine equal angles and arcs about the axis A as shown, these arcs being equal those of B, viz.: $mn = no$, etc., $= ac = de$, etc. Draw extended radii from the centers of these arcs as shown. Then the final radii for the finished wheels may be found by graphic proportion, by aid of the triangle ABO, as explained in Fig. 70. Thus make $Bh = Bb$, hs parallel to BO, us parallel to AO, and draw the line Ost. Then make $Ar = At$, and $Bp = Bt$, and we have two points r and

p in the final peripheries of the wheels A and B. The other points are found in a similar manner, as explained in Fig. 69, and the final wheels are drawn in by tracing lines through the points thus found.

The curves thus found, it is to be understood, are to serve as the pitch lines of a pair of gear wheels. Figs. 71 and 72 are examples of completed wheels by this process.

The shaper in practice will usually have the crank pin adjustable to different radial distances. In this case the above solution should be made for the crank at some mean position, where it is likely to be most used, as the law will be slightly deviated from for a different position of the pin.

2d. **A Bobbin Winding Problem.**—This problem, stated above, of winding yarn on a conical bobbin, or "cop," while feeding on the yarn spirally up and down the cone, and varying the speed of revolution of cone so as to draw the yarn at a uniform rate from a reel to avoid varied tension of yarn while winding, is believed to possess sufficient interest as an application of Fig. 69 to warrant its practical solution here.

Fig. 76 represents the conical bobbin with a spindle-like extention, showing also the cop of yarn in section.

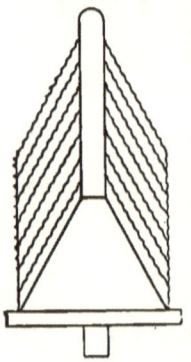

Fig. 76.

This is said to be the best form of bobbin and cop from which to supply yarn to knitting machines, since the yarn will lift off, and upward, from the position of Fig. 76, with the slightest and with the most uniform resistance—conditions of utmost importance to securing a smooth knitted web.

To preserve the same cone form of cop from end to end, the yarn must be fed upon the cone in a spiral or screw of uniform pitch, from base to spindle at vertex, and *vice versa*. The number of convolutions winding up, and the number back again, is arbitrary, and perhaps should be 20 to 40. But to simplify the present illustrative example, take it four; that is, four turns of the bobbin in winding from base to spindle, and four turns in winding back, eight turns in all. This multiplying of the motion, 8 to 1, may be done by using circular gearing between the non-circular gear wheel B and the bobbin, causing the bobbin to turn eight times as fast as B. This gearing, as by a screw of uniform pitch, is to work the feeding eye through which the yarn is to pass in being guided upon the cone.

Developing the cone with a single layer of thread upon it, we have the sector ODE, Fig. 77. DG represents one convolution of yarn, KN another, LP another, and IJ the last. These are Archimedean arcs, and may be put in a continuous spiral by shifting the second convolution KN around the center O till KN falls at GH, LP at HI, meeting IJ in the case the cone is so assumed, as here, that DE is the third of the circle. The whole thread in winding down the cone is then $DGHIJ$, and in winding back is the same taken in the inverse order.

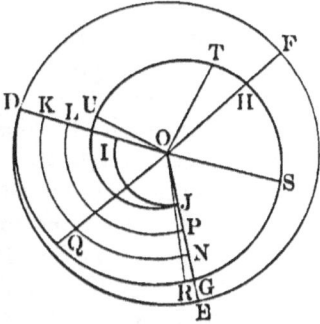

Fig. 77.

Now divide this thread into a certain number of equal parts, DQ, QR, RS, ST, TU, UJ, six in all. These portions must be wound upon the bobbin in equal times, in order to take the thread uniformly from the reel as required. Therefore the angles DOQ,

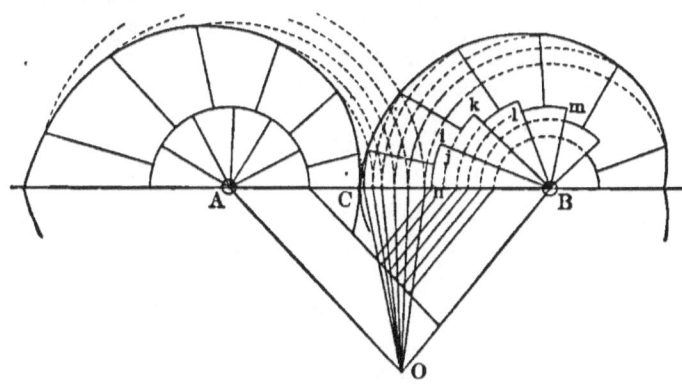

Fig. 78.

QOR, ROS, etc., must be passed in equal times by the finished wheel B, while the wheel A moves through equal angles in the same equal times.

In Fig. 78 lay off the six angles hBi, jBk, lBm, etc., as shown, varying in the same proportion as do the angles DOQ, QOR, ROS, etc., in Fig. 77; taking care that the six angles in Fig. 78, added together, equal 180°. For this result we have the added angles in Fig. 77 equal 360° + 120° = 480°; and hence multiply the angle DOQ by $\frac{180}{480} = \frac{3}{8}$, giving the angle hBi, Fig. 78. Also multiply angle QOR, Fig. 77, by $\frac{3}{8}$, giving the angle jBk, etc.

Then draw in circle arcs hi, jk, lm, etc., equaling corresponding arcs in A, and proceed as in Fig. 73 as shown, completing the half-wheels A and B, for which all the work is given in Fig. 78. The other halves are the same, A being symmetrical with respect to the line AC as an axis of symmetry, and likewise for B.

These wheels are one-lobed; A, the driver, differing somewhat from B in shape; and as constructed carefully in metal by above process, are shown in Fig. 73 for the same problem.

Motion of Driver A, a Variable.—In all the above problems, the velocity of the driver A has been supposed to be constant, but the solution is readily made where its velocity is variable, for which case it is only necessary to make the series of velocity sectors unequal to suit the varied motion, as was done for B, instead of making them equal.

Case of Multilobed Wheels.—In the above wheels, both driver and driven have been treated as unilobed, but multilobes are possible, particularly for the driver. For this, for a bilobed wheel A, it is only necessary to put into the scheme of angles for A twice as many velocity-angles as for B, three times as many for a trilobed, etc.; and similarly for a multilobed wheel B.

III. SIMILAR WHEELS.

1st. *Case of Similar and Equal Unilobed Wheels from Assumed Auxiliary Angles.*—In Fig. 79, A and B are centers of mo-

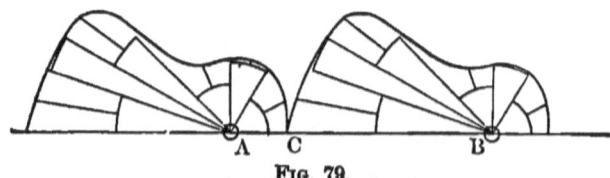

FIG. 79.

tion of the wheels. About A draw a series of auxiliary sectors, and about B the same series of sectors. Then, according to Fig. 69, draw equal arcs in all the sectors, and extended radii from the centers of the arcs as shown. Then the proportional radii, as of the auxiliary sectors and of the perimeters, may be found, when the wheels may be drawn in according to Fig. 70.

Fig. 79 shows the halves of a pair of wheels. These may be copied for the opposite halves, when the wheels will each be symmetrical; or another set of different auxiliary sectors may be drawn below AB and unsymmetrical similar wheels produced.

Fig. 129 is an example of a pair of wheels of this kind in metal.

2d. *Similar and Equal Multilobed Wheels.*—In Fig. 80 we have the half of a 3-lobed wheel where the lobes are all similar to each other, and one wheel similar and equal to the other, but where the lobes are unsymmetrical.

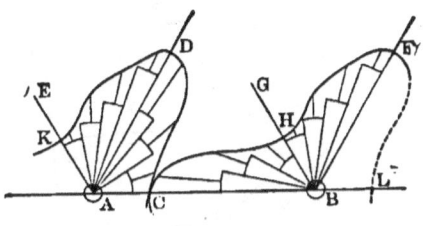

FIG. 80.

The wheels are arrived at by assuming a system of auxiliary sectors in the 60° angle CAD, and a like system in the 60° angle CBG. Determining the proportional radii between sectors and perimetric points as before, and tracing in the outline curves DC and HC, we find them similar.

Again, assuming a new set of auxiliary sectors, one like the other, in the 60° angles DAK and FBH, determining proportional radii, and drawing in the perimetric curves DK and FH, we will find completed a third of the wheel A; also a third of the wheel B.

By repeating these perimetric curves around through the remaining 60° angles of the wheels, we complete the pair of equal and similar 3-lobed wheels.

Thus equal and similar wheels of any number of lobes can be drawn.

3d. *Dissimilar Multilobed Wheels.*—Each wheel of Fig. 80 has lobes that are similar to each other, but they are evidently not necessarily made so, because each pair of divisional angles CAD and CBH are here equal to each other, but not necessarily equal to the next pair of divisional angles; and besides, the systems of auxiliary angles and sectors in one pair of divisional angles are not required to be the same as in any other pair.

4th. *Multilobes of Unequal Numbers of Lobes.*—In Fig. 81 we have a 3-lobed wheel mating with a 2-lobed one, where the

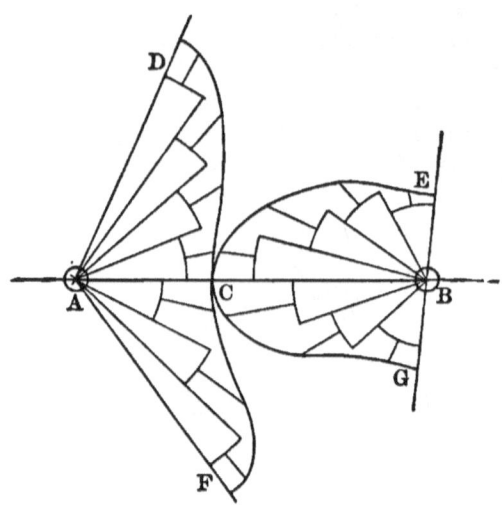

Fig. 81.

half-lobe CAD of A mates with the half-lobe CBE of B; and the half-lobe CAF with CBG. The sector $DAFC$ is the third part of A, and $EBGC$ the half of B.

In a similar manner, A may have any assumed number of lobes, and likewise for B.

In Fig. 81 the angle of the half-lobe $CAD = CAF$, and also $CBE = CBG$, but they may be unequal, though it would seem advisable for good results that the half-lobe angles have the relation

$$\frac{CAF}{CAD} = \frac{CBG}{CBE},$$

and that adjacent pairs of auxiliary angles in adjacent half-lobes be not greatly different in ratio of radii, that the perimetric curves may enter upon each other at the limits of sectors.

But it is plain that the auxiliary angles in any half-lobe are entirely arbitrary except at the lobe limits as above stated. In this way the auxiliary angles of a half-lobe may be changed at pleasure, thus modifying the perimetric arcs to suit fancy or requirements; it being only needful to note that the half-lobes that come to mate together be related in principle, as for the case where the lobes in one wheel are even and in the other odd.

CHAPTER IV.

CASE II. AXES MEETING.

SPECIAL BEVEL NON-CIRCULAR WHEELS.

It is usually advisable to refer wheels of this kind to spherical surfaces which are *normal* to all the elements of contact of the rolling cones whose vertices are at the center of the sphere and hence called *normal spheres*; the results of practical problems in drawing these wheels being either curves upon actual normal spheres, or ordinates, angles, etc., by which the curves may be traced upon the sphere by the pattern-maker as he follows the draftsman in the construction of these wheels.

Following the same order here as in plane wheels, we have—

FIRST. THE EQUIANGULAR SPIRAL.

1st. One-Lobed Wheels.—Take Fig. 82 to represent two views of the normal sphere on which a pair of spherical equiangular spirals are drawn for a pair of wheels A and B. The outline of one wheel, A, is shown at $Cabc\ldots f$ and $C''a'b'c'\ldots f'$ in the two projections; CBk, $C'B'$ being the mate; a copy of the first one, ACf.

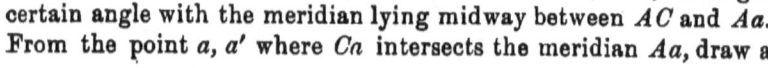

FIG. 82.

To draw this wheel, the actual normal sphere is to serve best as the drawing-board. From AA' draw a series of meridians at equal intervening angles of not far from 5° each. In Fig. 82 they are at 30° to avoid confusion of figure. We may assume a point C, C', making AC the shortest spherical radius of the proposed wheel. Then draw the line Ca, $C'a'$, making a certain angle with the meridian lying midway between AC and Aa. From the point a, a' where Ca intersects the meridian Aa, draw a

AXES MEETING. 63

line ab, $a'b'$, making the same angle, on the surface of the sphere, with the meridian midway between Aa and Ab that the line Ca did with the meridian between AC and Aa. This line intersects the meridian Ab at b. From b draw the line bc, $b'c'$, making the same angle with the meridian midway between Ab and AC as the previous lines did with their meridians.

So continue till the point f, f' is reached, giving a half-wheel of one lobe. The points $a'b'c'$, etc., may be symmetrically copied upon the other side of $C'A'f'$, thus completing the drawing of a one-lobed equiangular bevel non-circular wheel. This wheel may be copied at CBk for a mate, B, when we have the drawing of a pair of bevel non-circular wheels.

These drawings may, however, have been made upon normal spheres in the form of two separate spherical segmental shells in wood of uniform thickness, as shown in Fig. 83. With drawing completed, the pattern-maker could cut the wood to the finished wheels shown, following the drawing thus laid out for him. The drawing, however, as above described, is usually for the "pitch line" of a gear wheel, and the teeth may be laid out before going to the pattern-maker.

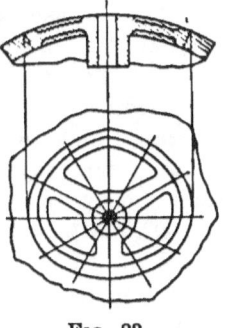

Fig. 83.

Where the mating wheels, as in this case, are both alike in pitch line, and perhaps for teeth as well, it is only necessary to complete one drawing and pattern to obtain castings for any number of pairs of cast wheels.

One point not to be overlooked is in regard to the size of the wheels as above explained. They may come out so as to require less or more than 90° between the axes, and for them to come out with any previously assumed angle of intersection of axes, several trials of tracing the line $Cabc\ldots f$ may be required. Any one of these trial curves, however, will work correctly with another copied from itself. The angle between the axes is the same angle as COf, which can be measured as soon as the curve Ccf is traced.

This curve of Fig. 82 differs from that of Fig. 95 of *Mechanical Movements*, of Professor C. W. MacCord, for the reason that there the lengths of the meridian-arc radii are made equal the radii of the plane equiangular spiral, which process makes the spherical spiral arc cut the meridians at a sharper angle for long than for short radii. This is most easily seen to be true in a quite eccentric 1-lobed wheel, in which, for short radii, the steepness of the curve on the sphere is nearly the same as for the plane spiral, while for the longer radii, nearly 90° of meridian arc, the distances from point to point are much less on the spherical than on the plane spiral, while the rate of change of meridian-arc radii remains the same as for the plane spiral. It would appear from this that the proposed wheels of Professor MacCord cannot work, because for both wheels of a pair the obliquity near the axis A or B differs from that for points more remote; while, as the wheels engage in action, the shorter radii of one wheel mate with the longer of the other where the obliquity is different.

2d. **Two-Lobed Wheels, Multilobed, etc.**—Having the curves of Fig. 82 drawn on the actual normal sphere, we may take the sectoral part $C'A'c'$ as the quarter of a two-lobed bevel non-circular wheel, a pair of which will work correctly. Or again, the sectoral part $c'A'f'$ is the quarter of a larger wheel, etc. With a spiral thus drawn on a sphere a wheel of any number of lobes can be made out, any one of which wheels will work with another like it.

That the axes of these wheels be at right angles, the wheels must be of certain size and may be made out by trial measurements on the spherical drawing. This tentative process, though seemingly tedious, is yet probably less troublesome than mathematical analysis, which in problems of mechanism often become tiresome.

3d. **Interchangeable Multilobed Bevel Non-Circular Wheels.**—Having drawn a spherical equiangular spiral as in Fig. 82, a sector of 180° may be selected from it for the half of a 1-lobed wheel. A second sector may be selected, of 90°, for the quarter of a 2-lobed wheel whose arc equals that of the 180° sector. Again, a third sector may be selected, of 60°, for the sixth part of a 3-lobed wheel and whose arc equals that of the 180° sector.

Any two of these wheels will work correctly together.

The sphere will put a final practical limit to the number in the series of these wheels, for a given spiral. Another spiral will furnish another series of interchangeable wheels.

Pairs of the above wheels selected to work together will have various angles between the axes, the point of intersection always corresponding with the center O of the sphere of Fig. 82.

Again sectors of different spirals may be combined in a lobe making unsymmetrical lobes.

Second. Elliptic Bevel Non-Circular Wheels.

1st. One-lobed Wheels.—This case is admirably treated by Mac-Cord in *Mechanical Movements*, Fig. 88, where spherical ellipses are shown as traced upon the normal sphere by means of two pins fixed at the focal points, around which is placed a loop of inelastic thread, when a pencil, drawing the thread tight and moved along against the tense thread, traces the "spherical ellipse" in a manner entirely similar to that so well known, of tracing the ellipse on a plane surface with two pins and a thread loop. Each ellipse has two focal points.

We may take this spherical ellipse as the base, and the center of the sphere as the vertex of an elliptic cone, a pair of which will work in true rolling contact when the vertices are at a common point, the center of the sphere, and the axes taken as lines radiating from the center of the sphere through the focal points above mentioned, one in each ellipse, and at an intervening angle equal that expressing the length of the major axis of the elliptic cone base; similarly as axes are taken at focal points A and B in Fig. 41.

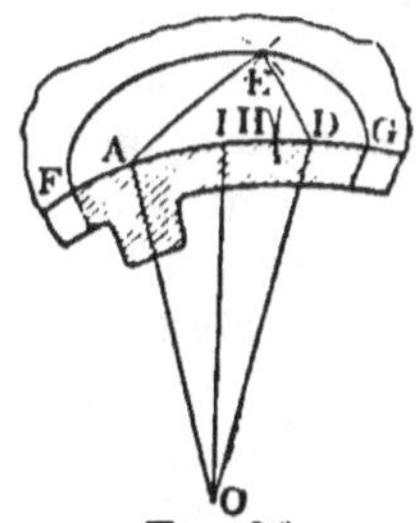

Fig. 84.

The wheels may be finished to the spherical form in the same way as illustrated in Fig. 83, the drawing being made on a wooden spherical blank or normal sphere, as drawing-board, as in Fig. 84.

On a great-circle arc of the sphere assume the portion $FADG$ as the major axis of the proposed spherical ellipse, making $FA = DG$, so that A and D may serve as focal points.

To find a point E in the spherical ellipse, take any arc FH in the dividers, and A with the focal point A as a center, describe an indefinite arc at E. Then with the remaining portion GH of the major axis, and with center at the focal point D, describe an arc at E cutting the former one, thus determining a point E in the spherical elliptic arc. Find other points in the same way sufficient for drawing in the curve GEF, the half of the elliptic curve required.

In order that the elliptic wheels may have axes at right angles, the angle FOG must be 90°; but the wheels and angles may be greater or less than this.

Taking the line AO as an axis, this wheel will mate with another like it.

Instead of spherical wheels for this case, Professor MacCord has shown that plane wheels may be obtained by cutting the elliptic cone by a plane at right angles to the line OI, giving an outline on the plane which will not be a true ellipse, though approximating one. The contact point of the pair of these wheels will always be found on a fixed *straight line* connecting the axes, and perpendicular to that which bisects the angle between the axes.

The advantage of the plane wheels over the spherical ones is doubtful, since in laying out the teeth on these blanks for gear wheels the teeth are not normal to the edges of the wheels except at four points, while for the spherical ones they are normal throughout.

As in the plane elliptical wheels of Fig. 41, a link may be mounted upon pins fixed at the focal points not occupied by the axes, as at D in Fig. 84 for wheel A. The pins at their bearings in the link must converge to the center of the sphere.

2d. **Multilobed Elliptic Bevel Wheels.**—These may be treated by applying the process of Figs. 85, 86, and 87 to Fig. 44, etc.

A pair of spherical elliptic unilobed wheels thus made have the same law of angular velocity as plane elliptic wheels of Fig. 44, as may be proved by aid of Figs. 85 and 86; while wheels of Fig. 84 have not.

Third and Fourth. Parabolic and Hyperbolic Bevel Non-Circular Wheels.

These can probably be best treated under the general case where any pair of plane wheels can readily be turned into bevels. See Figs. 85 to 88.

Fifth. Transformed Bevel Wheels.

Having given a pair of bevel non-circular wheels, in the form as traced on the normal spherical segments, the wheels being supposed divided into elementary pairs of angles, arcs, and radii, they may be transformed in the three ways mentioned for plane wheels, viz.:

1st. By multiplying any pair of elementary angles by the same quantity.

2d. By adding a constant length of meridian arc to all the meridional radii of one wheel, the elementary perimetric arcs remaining unchanged.

3d. By adding to one and taking an equal length away from the other of a pair of meridional radii, the perimetric arcs remaining unchanged in length.

Similar Bevel Multilobes.—Thus the half-wheels of Figs. 83 or 84 may be changed from 180° spherical sectoral wheels to 90° or 60°, etc., by Rule 1st, just stated, and several of them combined in one wheel of several lobes, two like ones of which will work truly together.

Dissimilar Interchangeable Bevel Multilobes.—By Rule 2d, Figs. 83 and 84 may be changed from 180° sectors to 90°, 60°, etc., where the perimetric arcs, as well as the difference of the limiting spherical radii, will remain unchanged in length.

In this way a 2-lobed, 3-lobed, 4-lobed, etc., wheel may be found from Fig. 83, which will mate with 83 or with each other.

Likewise for Fig. 84.

Partially Interchangeable Bevel Multilobes.—Starting with a pair of correct-working dissimilar bevel non-circular wheels as those of Fig. 78 turned into bevels, one of the pair may be transformed by Rule 2d into a series of spherical sectors of unchanged perimetric lengths, and difference of limiting radii; but with such angular widths as to add even to 360°; and likewise treating the other of the primitive pair, we have two series of wheels, any one of one series of which will work truly with any one of the other series, where no two of either series separately will work together.

Bevel Non-Circular Wheels of Combined Sectors.—Here, as in plane wheels, sectors of wheels of different class may be combined into one wheel, and the several mating sectors into another, the new wheels mating correctly.

Thus the half-wheel of Fig. 84 may be combined with the half-wheel of Fig. 83, the limiting radii being made to agree, and both constructed on equal normal spheres; the result being a pair of one-lobed wheels, one side being elliptic and the other the equiangular spiral.

Again, each of the above 180° sectors may be reduced to 90° by Rule 1st for transformation of wheels, or one to 70° and the other to 110°, etc., and combined, making pairs of 2-lobed wheels. Similarly, pairs of 3-lobed, 4-lobed, etc., may be brought out, but they will not be interchangeable.

Interchangeable Multilobes and Unsymmetrical Lobes.—Half-wheels of Figs. 83 and 84 may be transformed by Rule 2d into sectors that may be combined, both kinds into a single lobe, preserving the perimetric arcs and limiting radii, several of which lobes constitute wheels such as to realize interchangeable multilobes.

Velocity-Ratio in Bevel Non-Circular Wheels.—In Fig. 7 the velocity-ratio in circular bevel wheels equals the inverse ratio of the radii of the wheels, or ratio of the perpendiculars to the axes from the point of contact C. In this way in non-circular wheels we can find the velocity-ratio, or number of times faster one wheel revolves than the other at any instant. If the driver A revolves uniformly we can find the fastest and slowest motion for the driven wheel B; and that one is a certain number of times faster than the other is often all that is required, when the above wheels will answer the purpose.

CHAPTER V.

BEVEL NON-CIRCULAR WHEELS IN GENERAL.

The special cases of bevel non-circular wheels above considered were those of the equiangular spiral and of the elliptic curve, which could be readily studied in the bevel form, and for which cases the law of angular velocity was not important.

But in practical problems in mechanism where the *law* of motion of the driven wheel is essential, either for a part of or for the entire revolution of the driver, the forms of the wheels cannot be assumed to be certain spirals nor ellipses, but be determined both as mates, to suit the stated law of motion, as was the case in the like problem for plane wheels.

As the exact velocity-ratio depends upon perpendiculars upon the axes from the point of contact, and not upon meridian circle-arc radii, a complication enters by reason of which it is found advisable to work out the wheels as *plane wheels* first, and in conformity with the stated laws of motion, and then to convert them into bevels.

Therefore the wheels are first wrought out, as in Figs. 69, 75, or 78, as plane wheels, after which bevel wheels, possessing the same laws of motion, are determined.

Take Fig. 85 for the pair of plane wheels as worked out in strict accordance with the stated laws of motion, as in Fig. 69, A being the driver and B the follower.

Divide the wheels into parts or sectors of 5° or 10° each, the dividing lines serving as radii for reference.

In Fig. 86 draw the lines AO and BO with the same intervening angle AOB as that to be intercepted by the axes A and B of the finished bevel wheels. Draw A_1B perpendicular to AO, where its length is equal the line of centers AB of Fig. 85. Also draw B_1A perpendicular to the line BO, where its length is also equal the line of centers AB, Fig. 85. Then draw the line AB,

Fig. 86. By this construction the triangle ABO is isosceles. Also draw the circle $A'B'$ from O as a center, to represent the normal sphere in section.

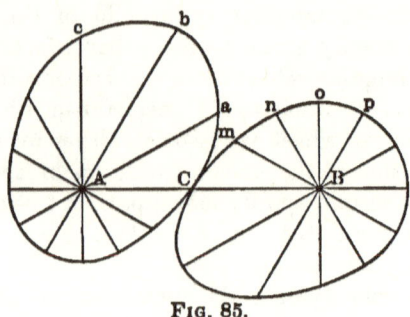

Fig. 85.

Then lay off perpendiculars AC, Aa, Ab, etc., Fig. 86, equal to the radii AC, Aa, Ab, etc., of wheel A, Fig. 85, and draw parallels

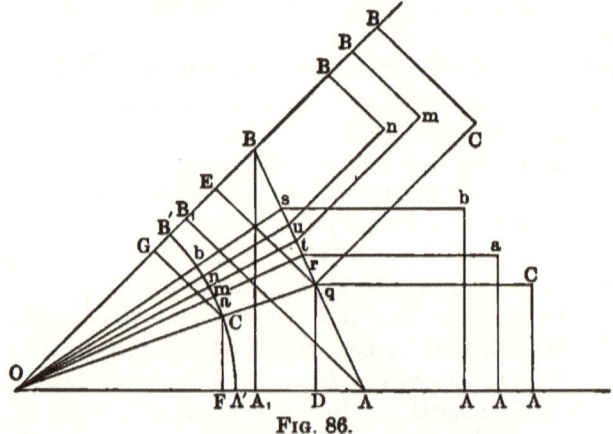

Fig. 86.

to AO, as Cq, ar, bs, etc., giving points of intersection q, r, s, etc., on AB. Likewise draw perpendiculars BC, Bm, Bn, etc., Fig. 86, equal to the radii BC, Bm, Bn, etc., of wheel B, Fig. 85, and draw parallels to BO, as Cq, mt, nu, etc., giving points of intersection q, t, u, etc., on AB.

Then draw the lines from the points q, r, s, t, u, etc., to O, determining the intersection points C, a, b, m, n, etc., on the circle arc $A'B'$, representing the spherical blank in section.

This completes the drawing preparatory to work on the spherical blank in wood.

Let Fig. 87 represent the spherical blank for a pattern for the wheel A in plan and section, the radius AO being equal $A'O$ of Fig. 86 of the normal sphere. On the plan draw meridian lines with the same scheme of intervening angles CAa, aAb, bAc, etc., as in wheel A, Fig. 85. Then with compasses opened to the arc $A'C$, Fig. 86, place one point at A, Fig. 87, and strike an arc at C, cutting the meridian at the point C as shown. Then the point C, Fig. 87, is one point in the periphery of the bevel wheel A.

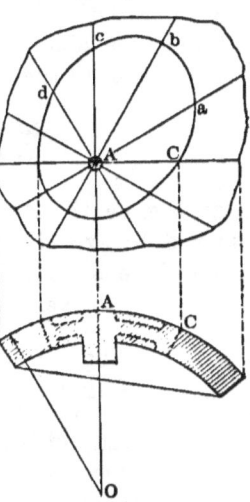

Fig. 87.

Another point is found by taking $A'a$, Fig. 86, in the dividers, and laying it off on the meridian Aa of Fig. 87, determining the point a of the periphery. Likewise, using distances from Fig. 87, determine points b, c, d, etc., on all the meridians of wheel blank A, Fig. 87, when the wheel outline may be drawn in and the blank cut down thereto.

The sectional view of Fig. 87 shows how the blank can be cut away within the outline to a spherical web, or to the desired number of arms, as for the pattern from which to obtain castings for wheel A.

A second spherical blank for wheel B is also turned up with the radius AO, Fig. 87, equal $A'O$, Fig. 86, as before, for wheel A.

On this blank are drawn meridian lines with the same scheme of intervening angles as shown on wheel B, Fig. 85. Also on all the corresponding lines of the blank lay off the arcs $B'm$, $B'n$, etc., as in case of wheel A; draw the perimetric outline, and cut the wood blank to shape as before, giving a wheel B to mate with wheel A, as a pair of patterns for castings in pairs, as the final result of the problem.

The wheel outlines thus found are of course usually the pitch lines for non-circular gears, the construction of teeth for which is explained later.

In the above the wheels were treated as one-lobed. But either or both may be of several lobes, or one of a pair may be internal.

Usually the angle AOB, Fig. 86, is a right angle, and $AB = AB$ (of Fig. 85) $\sqrt{2}$.

The proof of the above figures and process was not attempted with the explanation. To show it to be the correct one to carry forward into the finished bevel wheels the same law of angular motion as the plane wheels of Fig. 85 possess, we note that the velocity-ratio for the wheels, as shown in Fig. 85, is

$$\frac{\text{Angular velocity of } A}{\text{Angular velocity of } B} = \frac{BC}{AC}.$$

In Fig. 86, if OA and OB be axes, we find that for an arm AC attached to axis A, operating an arm BC attached to axis B, by having the ends joined together at q, with the arms suitably located on the axes, the velocity-ratio will be

$$\frac{\text{Angular velocity of } OA}{\text{Angular velocity of } BO} = \frac{Eq}{Dq} = \frac{BC}{AC} = \frac{GC}{FC}$$

= velocity-ratio of the finished wheels for contact C, since FC, Dq, etc., are all perpendicular to the axes of Figs. 86 and 87.

The same is true of all other pairs of radii. Hence the bevel non-circular wheels thus obtained have the same laws of angular motion as the plane wheels of Fig. 85, as required.

Fig. 88 is an example in wood for patterns for castings, of a

FIG. 88.   FIG. 89.

pair of bevel elliptic unilobed and bilobed wheels, laid out by the above process.

In Fig. 89 we have a photo-process copy of a pair of bevel non-circular wheels in metal, laid out by above process from Fig. 73, consequently possessing the same law of angular velocity. The wheels are about 7 inches between centers, and at 45° angle of intersection of axes.

LAYING OUT BEVEL NON-CIRCULAR WHEELS DIRECTLY ON THE NORMAL SPHERE.

1st. One Wheel Given, to Find its Mate.—Let $ACbD$ represent the given or assumed wheel traced free hand upon the normal sphere, the pair of wheels to be complete and not sectoral.

Assume B for a trial center for wheel B, and the arc ACB on a great circle of the sphere for a line of centers. Assume points a, b, c, etc., and find mating points m, n, o, etc., such that $Cm = Ca$, $mn = ab$, etc.; also spherical arc $Bm +$ arc $Aa =$ arc ACB; arc $Bn +$ arc Ab $=$ arc ACB, etc., when the perimetric curve for wheel B may be traced through the points C, m, n, o, etc.

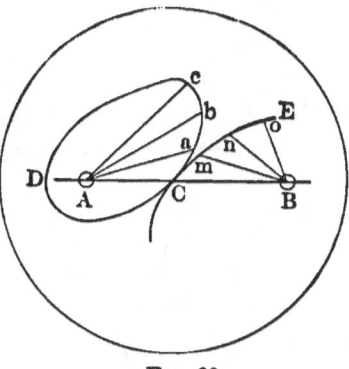

FIG. 90.

If B was assumed too far from A the wheel B will have a gap in it, when another trial point B may be assumed and the work repeated till the wheel B closes.

2d. Laws of Motion Given, to Find the Wheels.—Let Fig. 91 represent the actual normal sphere, and assume the wheel centers A and B on it or axial lines $A'O$ and $B'O$ at the desired intervening angle, say 90°.

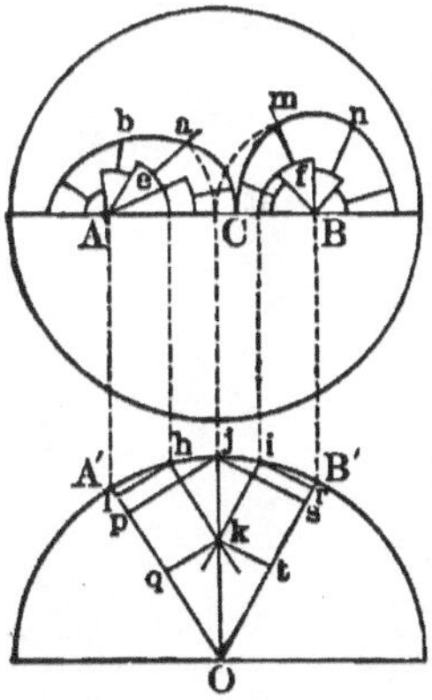

Lay out around A the auxiliary sectors with angles to represent the motion of A, the sectoral angles being passed over in the movement of the finished wheel A in equal times. Likewise draw the mating auxiliary sectors around B as shown.

To find a pair of mating points a and m in the peripheries of the finished wheels, make the arc $A'h = Ae$, and $B'i = Bf$; find k by parallels to $A'O$ and $B'O$, and draw Okj. Then make $Aa = A'j$, and $Bm = B'j$, giving a and m for the desired mating points.

FIG. 91.

Thus proceed with all the auxiliary sectors, when the wheels may be drawn in through the points a, b; m, n, etc.

This will make the velocity-ratio of the finished wheel the same as that of the auxiliary sectors rolling upon each other, because the velocity-ratio for the points a and m in contact at C is the same as for the sector Ae rolling on the sector Bf, or the same as for axis $A'O$ to drive axis $B'O$ by contact of ends of perpendiculars lh and ri; hence

$$\text{Velocity-ratio} = \frac{lh}{ri} = \frac{kq}{kt} = \frac{jp}{js},$$

so that the finished wheels have the desired velocity-ratio.

These wheels will each turn through a pair of mating angles in equal times.

If A revolves uniformly, as is usually the case, the sectors for A will be equal to each other.

Similar Wheels, Lobed Wheels, etc., may be worked out on the normal sphere, similarly as for plan wheels in Figs. 79, 80, and 81.

CHAPTER VI.

CASE III. AXES CROSSING WITHOUT MEETING. SKEW-BEVEL NON-CIRCULAR WHEELS.

For this we will attempt only the general case, or the counterpart of non-circular bevels as brought out in Figs. 85, 86, and 87, as pertaining to non-circular skew bevels; wooden blanks being here presumed also, as receiving part of the work of the draftsman.

Fig. 92 is a pair of circular hyperboloids answering for the case

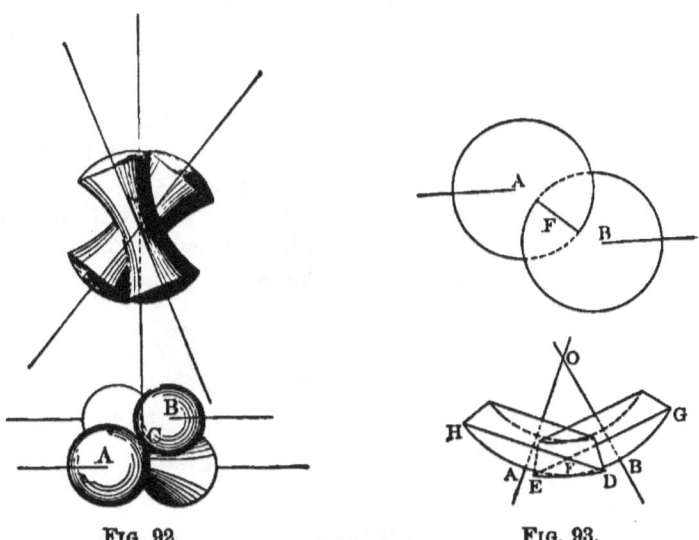

FIG. 92.   FIG. 93.

of pitch surfaces for circular skew-bevels, given to aid in acquainting the mind with the forms we have to study.

In the present case the rolling or pitch surfaces are non-circular hyperboloids with fixed axes, and a given shortest distance between them.

75

At C the surfaces do not coincide, but lie at an angle with each other so that if extended beyond C they would intersect, as in Fig. 93 at F. This figure may be taken to represent a pair of blanks from which we could cut a pair of non-circular skew-bevels. These should be turned in a lathe, of nearly spherical form, such that their line of intersection AFB as seen in plan is a circular line with O as its center.

The line AFB, in space, will be so nearly a circle that it will be assumed to be one, and a common meridional line of the convex surfaces of A and B, so that DAH and EBG will be equal spherical surfaces intersecting in the line AFB, made to a radius greater than AO to be determined.

The final result in the laying out of these wheels will be the non-circular outlines on the blanks A and B, to which the material may be cut, to realize the final rolling wheels.

As a first step in the process of obtaining these outlines, determine a set of non-circular wheels with axes parallel, as in Fig. 85, that will work in conformity with the laws of motion as required for the skew-bevels.

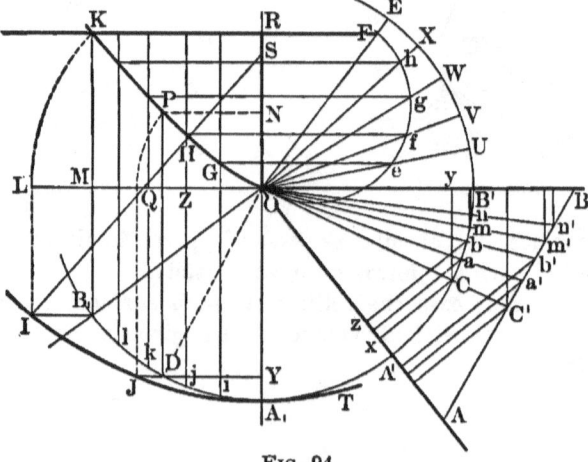

Fig. 94.

Also draw Fig. 86 from 85 as before, making the angle AOB equal the angle AOB of Fig. 93, as the given angle between the axes in the plan of the skew-bevels; and the circle $A'CB'$, Fig. 86, with the radius AO, Fig. 93.

Then draw Fig. 94, where the portion AOB is the same as Fig. 86, and where OA_1 and OB_1 is the horizontal projection and KR and O the vertical projection of the axes, LO being the ground-line.

In the vertical projection the common perpendicular between the axes is OR, projected in plan in the point O.

The line of contact between the rolling surfaces is here, in elevation, a shifting line PN, always parallel to KR and LO, and in plan it is a shifting radial line OD, PN and OD being the projections of that line as in one position.

The circle A_1DB_1 is taken as the projection of the line of intersection AFB of the extended wheel-blanks shown in Fig. 93, A_1O being the distance of the pole of wheel A from the common perpendicular between the axes, B_1O being the like distance for wheel B; which distances are equal $A'O$ and $B'O$.

The line of intersection A_1DB_1 of the extended wheel-blanks is seen in elevation in the line OPK.

To correctly draw the line, divide the arc A_1DB_1 into equal parts by points i, j, k, l, etc.; project lines up from these points to KR; then extend KR to F, where the angle $FOB' = $ angle A_1OB_1; draw the semicircle FeO, which intersect at e, f, g, etc., by radial lines from O to points of equal division in the arc $B'UVE$, and project lines from the points e, f, g, etc., over to intersect the lines previously projected up from A_1B_1 to KR. Through the proper points of intersection G, H, etc., draw the line $KPHGO$ for the correct line of intersection sought. This may be regarded as a line on a vertical circular cylinder A_1DB_1 with center at O, which, developed, would show the line OPK as slightly S-shaped.

In the geometrical work this line will be used in its true form; but for turning up the wheel-blanks an approximating circle arc will be used, found by revolving OPK around axis O to the horizontal plane when the point K describes the arc KL, falling at L, and B_1 falls at I. To draw the circle arc A_1JI a center S is found on the line A_1O extended, in such position that $A_1S = IS$.

The circle IA_1T is a meridian arc to which the non-circular skew wheel-blanks are to be turned in the lathe in order that their line of intersections, as in Fig. 93, may be such as to run through AFB, as required.

In Fig. 95 we have the projection of the wheel-blank A turned up to fit a circle pattern cut to the circle IJA_1T, Fig. 94. On this blank lay off the angles CAa, aAb, etc., as a copy of the angles similarly lettered on the drawing of the plane wheel, Fig. 85, previously explained, drawing the radiating lines or meridians to the edge of the blank, as shown in Fig. 95.

To find the point C on the wheel-blank radius AC, take the length Cx, Fig. 94, in the dividers, and, without changing, place one foot at P on the curve OPK at such position that the other foot falls at N on the line OR, making PN perpendicular to OR.

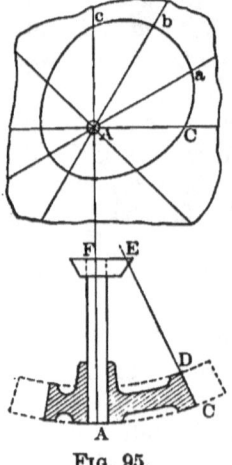

Fig. 95.

Then, retaining the one foot of the dividers at P, swing the dividers, and open them till the other foot falls at O. Retaining this length in the dividers, place one foot at J on the circle $A_1 JI$ while the other foot falls on $A_1 O$ at Y, making $JY = OP$ in the diagram and perpendicular to $A_1 O$. Then swing the dividers around J and open them till the other foot falls at the point A_1. With this length $A_1 J$ in the dividers lay off AC, Fig. 95, on the meridian AC of the wheel blank A.

To find the point a, Fig. 95, take the distance az, Fig. 94, in the dividers, and apply it on another line PN, PO, JY, JA_1, as before, and finally Aa, Fig. 95, and so proceed for all the points in the periphery of A, when the outline of the wheel A may be drawn in.

Otherwise explained, lay off Cx, Fig. 94, from R on RK. Project the end of this line down to P. Revolve P to Q about O. Project Q to J. Then $A_1 J = AC$ on wheel-blank Fig. 95. Proceed likewise for other points.

For wheel B, as for A in Fig. 95, take distances from C, m, n, etc., to line OB', Fig. 94, and apply these as before from R toward K. Project to P, revolve about O to Q; project Q to J, and take JA_1 to apply on wheel-blank of wheel B in the manner indicated in Fig. 95 for meridian distances BC, Bm, Bn, etc.

When ready to cut the edge of wheels A and B to the perimetric lines just laid out, a question will arise as to what direction CD or bevel to give across the edge of the wheel, as it will not always be normal to the spherical outer and inner surfaces, and the right-line elements of the surface CD will be askew with the sphere normals.

One way to proceed is to cut the minimum non-circular section of the wheel from a piece of plane thin material, or thicker with the edge bevelled, as EF, Fig. 95, and to mount this upon an axial rod, FA, extended from the wheel blank A, Fig. 95, making $AF = OA_1$, Fig. 94, or OA, Fig. 93.

To lay out the small wheel EF first draw a series of diametric lines with the intervening angles the same as in the upper part of Fig. 95, or in the plane wheels of Fig. 85. The lengths of the radii will be the several heights ON, Fig. 94, which may be noted as the lengths Cx, az, etc., are placed in the several positions PN as explained. These heights ON are to be laid off on the proper radii of EF, Fig. 95, in order; when the outline may be traced in, and the section EF cut to shape. The radial lines on EF, as well as on A, should be plainly marked and lettered, or numbered, to distinguish them. In mounting EF upon the rod the mean radius approximately should be selected and noted, as, for instance, that corresponding with AC in the figures, and this radius on EF placed at an angle with the corresponding radius on wheel, which angle is PON, Fig. 94.

Then in dressing off the edge of wheel A, Fig. 95, use a thin straight-edge, or thread across from EF to AC, trained upon corresponding points of the two, cutting through from C to D till it coincides with the straightened thread stretched from E to C, and likewise for the several points of radii noted, when the intermediate portions can be cut by eye.

Wheel B is to be treated in a similar manner.

This gives us the rolling or pitch surfaces of the wheels.

Instead of using the small wheel EF, Fig. 84, as a guide in stretching the thread to use while trimming off the material at CD and other points, it may be advisable to use a second blank like AC, turned to the same spherical surface except it may be thinner by cutting away on the concave side. Let us call this blank M.

On the spherical convex side of this second blank M we are to lay out meridians, straight or slightly curved, as explained for Fig. 95, but with intervening angles in the inverse order of those of Fig. 84. On these meridians the same radial distances AC, Aa, etc., are to be laid off also in the inverse order.

This wheel, cut to line with quite a relief bevel, is to be mounted upon an axial rod like AF, Fig. 95, extended to twice the distance AF, or to twice OA_1, Fig. 94; and at the intervening twist angle between corresponding radii of twice PON, Fig. 94; and with the concave sides of the blanks toward each other, and twisted in the right direction. See Fig. 139. Then in dressing off the blank at and along CD attach the line to the opposite blank M, at the corresponding point C, and draw the line straight, and apply to CD as often as desired while dressing away the material, until coincidence is obtained entirely across CD, making this one right-lined element of the surface. If we call this element CC, because connecting the points C and C of the blanks, the next point may be aa, and the next bb, etc. See Fig. 95.

After dressing the blank across at CD, move the line and dressing action to the next point aa, finishing which as before; move again to bb, etc., for the whole periphery of A. On the portions of surface of the edge of A between the line elements CD, etc., as above found, the material may be cut away by eye, or if preferred the line may be stretched to intermediate points.

Proof for the construction of the line OPK, Fig. 94, as described, is found in the formula expressing the proper division of the common perpendicular OR, Fig. 94, or $A'B'$, Fig. 10, into segments or gorge radii, in accordance with the velocity-ratio, viz., Fig. 10, calling angle $A'C'O' = \alpha$, $B'C'O' = \beta$, $A'C'B' = a$, and the common perpendicular $= c$.

$$O'C' \tan \alpha = A'O' = x; \quad O'C' \tan \beta = B'O' = y;$$

whence $\dfrac{\tan \alpha}{\tan \beta} = \dfrac{x}{y}$; also $x + y = c$ and $\alpha + \beta = a$.

Eliminating y and β, we get $\dfrac{\tan (a - \alpha)}{\tan \alpha} = \dfrac{c - x}{x}$,

which may be reduced to $x = c \dfrac{\sin \alpha}{\sin a} \cos (a - \alpha)$,

in which, for Fig. 94, $c = OR$, $a = AOB = A_1OB_1 = BOE$; $\alpha = $ any angle, as $B'OV$, and $x = $ the corresponding height HZ.

Thus $OF = OR$ secant $ROF = c \cdot \operatorname{cosec} a = \dfrac{c}{\sin a}$;

$Of = OF \cos VOE = \dfrac{c}{\sin a} \cos (a - \alpha)$;

$HZ = OF \sin B'OV = \dfrac{c}{\sin a} \cos (a - \alpha) \sin \alpha = x$.

Hence it appears that the height HZ, Fig. 94, is the result of the graphic construction of the above formula, and that the position of the point of contact or intersection of the line of contact with the line OP is correctly determined for each set of radii of the plane wheels, and that consequently the radii of the wheel blanks, made out as described, are correct.

To be strictly correct, instead of meridians AC, Aa, Ab, etc., Fig. 95, as mentioned, those lines as seen in projection should be laid off as copies of the line OPK as seen in projection in Fig. 94, and at intervening angles, the same as before. Then FE, Fig. 95, should be located with respect to a particular point P and angle PON.

CHAPTER VII.

NON-CIRCULAR WHEELS FOR INTERMITTENT MOTION.

These wheels might properly have been considered under the cases of axes parallel, axes meeting, and axes not parallel and not meeting, but as their peculiarities in other respects are greater, it was thought advisable to treat them together, in deference to that refinement of classification.

They are here classified in two ways, with reference to the devices for stopping and starting the driven wheel, viz.:

1st. Where easement spurs are used together with locking arc of greater radius than that of the rolling arc; and

2d. Where the locking arc is smaller than the rolling arc, with easements upon the ends of the pitch lines.

1st. Where Easement Spurs are Employed.

In Fig. 96 we have an example of the first kind, where easement spurs or prongs are used for stopping and starting the driven wheel with reduced shocks. Here the arc DEF is non-circular and rolls correctly on the arc GMH, the arc from near D and F through G and H being circular and serving simply as a lock for wheel B while standing still during its period of rest or intermitted movement.

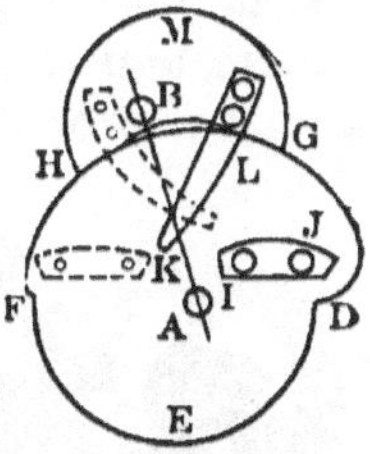

The curved piece IJ fast upon A, and KL fast upon B, have initial contact at the points I and K; when A is rotated so that I approaches K, and when I is carried down near to A, the quickness with which B is started is reduced, as well as the shock due to it. Hence for rapid motions AI should be short, while for slower speeds it may be greater. In some cases IJ may be a mere pin, as in Fig. 11 for circular rolling curves.

Fig. 96.

In either case the shape of DG and saddle curve GH should work with but little backlash when IJ and KL are in running contact.

The spurs here, as well as in Fig. 12, may work by rolling

82 PRINCIPLES OF MECHANISM.

or sliding contact, though the latter will be found most convenient to lay out, and no more objectionable, unless it be in the matter of lubrication.

Fig. 97.

2d. Where Locking Arcs and Easements are Employed.

In Fig. 97*t* the wheels have the locking arc of smaller radius than the rolling arc DEF, instead of larger as in Fig. 96. The shapes of D and F will be much the same as in Fig. 15.

The rolling arcs in Fig. 97, as well as in Fig. 96, may have any non-circular form desired, according to requirement of practical problems.

As in circular wheels, these may be made with axes meeting or axes crossing and not meeting, according to principles laid down in the proper places as pertaining to these problems.

Fig. 98 is a photo-process copy of a pair of wheels like Fig. 96, turned into bevels, with axes meeting at 60°, 7 inches between centers, and of very considerable eccentricity, as best shown in the second part of Fig. 98.

Fig. 98.

The process of Figs. 85 to 87 was followed in laying out these wheels.

DIRECTIONAL RELATION CHANGING. VELOCITY-RATIO CONSTANT OR VARYING.

ALTERNATE MOTIONS BY ROLLING CONTACT.

Pitch Lines for Alternate-Motion Gearing.

Among these movements there are those that are **necessarily** limited in extent of movement, and others not.

First. Limited Alternate Motions.

Velocity-ratio Constant.

In Fig. 99 the driving half-wheel A turns continuously in one direction while the driven piece B reciprocates, and hence the change in directional relation.

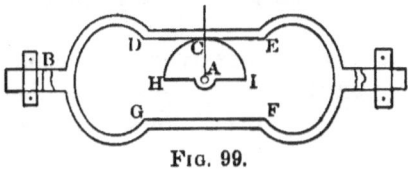

Fig. 99.

In the present case the half-wheel is circular, and rolls on the straight line DE for movement in the one direction, the line DCE being equal the semicircumference HCI. The arrangement is to be such that as the driven piece B, in moving to the right, reaches that position where H comes to contact with the extremity D, the other end I of the half-wheel arc is just on the point of taking contact at G, so that by further motion of A the rolling is transferred from DE to FG and so continues with B moving to the left, till the last point, H, of the wheel makes contact at F, when the contact is retransferred to DE, and the motion again reversed.

It is plain that the greatest extent of movement of B is the length $DE = GF = HCI$, and hence the movement is termed "limited."

Also it is plain that if A revolves uniformly, the motion of B will be uniform and the velocity-ratio constant.

Fig. 99 is simply the theoretical representation of the movement. In practice, gear teeth, as well as suitable positive engaging and disengaging devices, are to be applied, by which the movement will be still further limited, which gearing and devices will be treated under the proper headings.

Velocity-ratio Varying.

In Fig. 100 the driver A is a 180° sector of a log. spiral and, as before, rotates continually in the same direction, while the point of contact shifts over from the straight inclined line DCE to the parallel line FG, simultaneously with which the direction of motion of B is reversed.

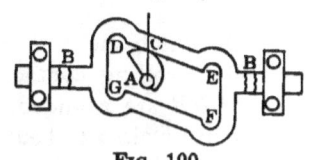

Fig. 100.

If A revolves with uniform velocity, the movement of B is hastened as the contact at *C* approaches *D* or *F*, so that the velocity-ratio is a varying quantity.

Drawing a line from the center of *A* through *C* in a direction perpendicular to the slide bearings *BB*, we have the line of centers, the center of motion of *B* being at an infinite distance from *A*, since the line *DC* is an arc of a log. spiral with the pole at infinity.

The velocity-ratio is the same as given in connection with Fig. 27, the rate of motion of *B* being the same as that of the point *C* as if revolving about *A* with the angular velocity of *A*.

Another example of velocity-ratio varying is given in Fig. 101, where *A* may be taken as a portion of any one-lobed wheel, and the curve *DE* or *FG* found such as will mate with it. The same formula for velocity-ratio applies as in Fig. 100 or 27.

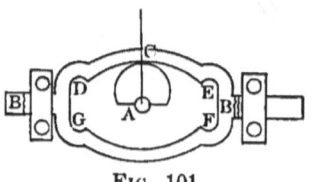

Fig. 101.

SECOND. UNLIMITED ALTERNATE MOTIONS.

Velocity-ratio Constant.

The Mangle Rack.

In Fig. 102 *A* acts by rolling contact against the bar *BB* secured to the flat surface of *DE*, in which latter a groove, *FG*, is

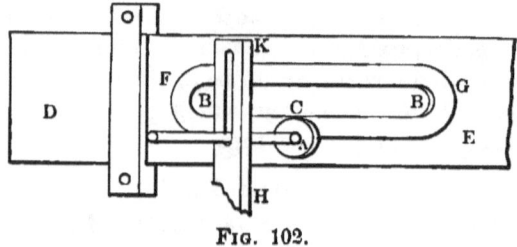

Fig. 102.

cut into which the end of the shaft of *A* is inserted, that *A* may be held in rolling contact with the barlike raised part *BB*. The ends of *BB* are rounded, equidistant from which the groove *FG* follows in its circuit about *BB*. The piece *DE* is fitted to slide in straight guides, so that as wheel *A* continues to revolve in the same direction, the bar *BB* and attached slide *DE* will be moved till A reaches the extremity of B, when further rotation of A will cause A to pass around upward at one end and downward at the other end, in continued revolving of A and reciprocation of BB and DE.

A bar HK has a slot through which the shaft of A passes to prevent it from swinging to the right or left.

This mangle-rack movement may be given a piece BB of any length, without limit. The part BB may be narrow, even reduced to a mere line.

The velocity-ratio is the same as that in Fig. 27 or 100.

The Mangle Wheel.

In Fig. 103 we have much the same construction as in Fig. 102, except here BB and FG are formed in a circle around O, the part DE being here a circular disk mounted on a center at O, so that the driven part BB and DE has an alternating circular motion.

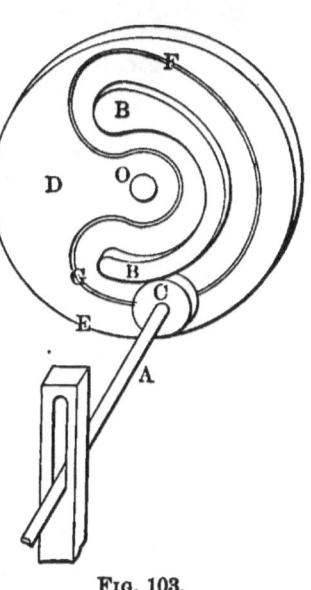

Fig. 103.

The velocity-ratio is constant while B is moving in one direction, where BB is circular about O also in the other, though in the second case it differs from that of the first when BB has a sensible radial thickness.

Velocity-ratio Variable.

The velocity-ratio will be variable when the piece BB in Fig. 102 is curved on a straight slide, or straight on a curved slide, or with varied width.

For Fig. 103 the piece BB may be of varied width, or non-circular with respect to the centre O, or in form of a spiral, even reaching to several turns about O, and with A non-circular, combined with the above modifications.

AXES MEETING.

Bevel wheels in these movements are practical, and probably skew-bevels also, the construction of which must have due regard to the principles of bevels, or skew-bevels, as laid down under rolling contact.

In Fig. 104 is given a photo-process copy from a pair of mangle wheels under Axes Meeting at a right angle. The pitch line is merely a circle on a cylinder set with teeth, which are simply a row of pins with which the driver engages on either side.

Fig. 104.

PART II.

TRANSMISSION OF MOTION BY SLIDING CONTACT.

CHAPTER VIII.

SLIDING CONTACT IN GENERAL.

In elementary combinations of mechanism acting by sliding contact, there are a driver and a follower with axes at a fixed distance apart, the driver acting upon the follower at a point of contact at some distance to one side of the line of centers, in consequence of which there will be sliding of one piece against the other at the contact point, since rolling action occurs only when the contact is on the line of centers.

In thick pieces the contact will be along a line of tangency, which line must be continually parallel to the axes when the latter are parallel, or meet with them at a common point when they are meeting. In each of the two pieces, all sections normal to the axis are similar figures.

Velocity-ratio in Sliding Contact.

In Fig. 105, take A as the center of rotation of a piece $AJDK$, with contact at D against the piece $BLDM$. Then when the piece AD is turned through some angle toward BD, the latter must turn also, when a sliding action will occur at the contact D. Hence AD may drive BD by sliding contact at D.

To find the *velocity-ratio* for this turning of the pieces AD and BD about their centers A and B, while thus transmitting motion by sliding contact, find the *center of curvature* of the piece $AJDK$

or of the curve *JDK* at *D*, as some point *E*. Also find the center of curvature *F*, of the curve *LDM* at *D*, for the piece *BLDM*.

Then *DE* will be the *radius of curvature* at *D* of the curve *JDK*, and *DF* the *radius of curvature* at *D* of the curve *LDM*; and pieces might be cut from thin wood, one of shape *AJDK*, and the other of shape *BLDM*, which, when held by pins at *A* and *B* for axes, could turn by sliding upon each other in the manner proposed. For a small amount of movement, sufficient to determine the velocity-ratio, the distance *EF* will be constant, it being the sum of the radii of the curves in contact at *D*. We observe that in this case *E* is a fixed point on the piece *AD*, and *F* a fixed

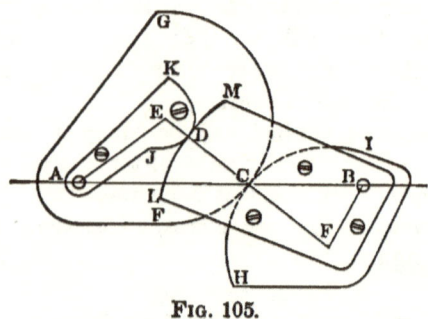

Fig. 105.

point on the piece *BD*. Now cut another pair of pieces, *AFCG* and *BHCI*, making *E* the center of curvature of the curve *FCG* at *C*, and *F* the center of curvature of the curve *HCI* at *C*, the point *C* being in this case chosen where the line *EF* intersects the line of centers *AB*.

If now the two pieces *AEC* and *AED* are fastened together with a common axis at *A* and coincident at *E*, likewise if the pieces *BFC* and *BFD* be united with a common axis *B* and coincident at *F*, and the two be mated on their centers *A* and *B* by axial pins and brought to running contact, then for a small movement the pieces will be in rolling contact at *C*, as in Fig. 2, because this point is on the line of centers: and in sliding contact at *D*, as at first proposed; and the two actions will have a common velocity-ratio which equals $\frac{BC}{AC}$, as in Fig. 2.

Therefore, to determine the velocity-ratio of a pair of pieces *AJDK* and *BLDM* transmitting motion by sliding contact, draw a common normal *DC* to the curves in contact, extending it to intersect the line of centers; then the velocity-ratio is given by the equality

$$\frac{V}{v} = \frac{BC}{AC},$$

where V is the angular velocity of A, and v the angular velocity of B. Hence the velocity-ratio in sliding contact is the inverse ratio of the segments of the line of centers as the latter is cut by the common normal to the sliding curves.

In Fig. 105 only a small movement was supposed to occur, because the curves in common tangency at D and C may not be circular to any considerable extent. But to realize an indefinite movement under like conditions, it is only necessary to suppose the curves to be so shaped that the point of tangency C shall remain on the line of centers as for the non-circular wheels of Fig. 51, while the common normal EDF shall continue to pass through the same tangency point C. For this, the centers of curvature E and F may be continually changing positions.

The curves FCG and HCI may be circles to the centers A and B, as in the pitch lines of circular gearing, when the point C of the common normal CD becomes stationary on AB.

SLIDING AND ROLLING CURVES WITH COMMON LAW OF MOVEMENT NOT REQUIRED TO BE CONCENTRIC.

In Fig. 106 let the curves $AaCc$ and $BbCd$ be a pair of correct-working non-circular rolling wheels in rolling contact at C; and

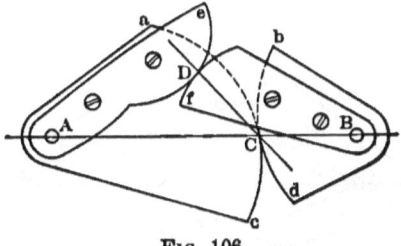

Fig. 106.

ADe and BDf a pair of pieces in action by sliding contact at D, and of such shape that the common normal at D, prolonged, shall continually meet the point C. Then, because the velocity-ratio of the non-circular wheels is BC over AC, and because the velocity-ratio of the sliding curves in contact at D is also BC over AC, it follows that DC must be a common normal at D, but not necessarily at C; and that the law of the velocity-ratio is common to both actions so long as the curves are so shaped that DC is a common normal at D and meets the point C.

Fig. 107 is a photo-process copy of a model in metal of what is shown in Fig. 106. The lower sectors are the rolling arcs, and the

FIG. 107.

upper the sliding arcs. Their lengths are such that they all begin and end action together. The sliding curves cross the rolling curves as in Fig. 108.

SLIDING CURVES TO WORK CONFORMABLY WITH ROLLING CURVES.

In Fig. 108 let $AlCjck$ and $BoCmgn$ represent a pair, or por-

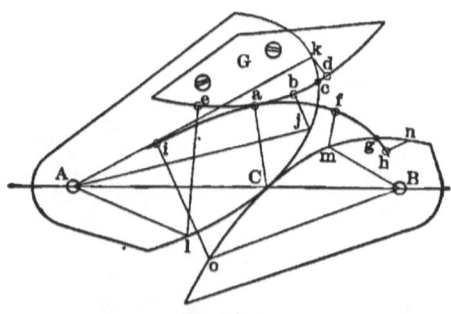

FIG. 108.

tions of a pair, of correct-working rolling curves centered at A and B. Assume a sliding curve $eabcd$, which may be a templet piece G

cut to suit, and made fast to wheel A, the latter being also a templet piece as shown to mate with a templet piece B.

Find the mating curve $iafgh$, which, by sliding action upon $eabcd$, will transmit the same motion, A to B, as by the mutual rolling of the rolling curves or wheels A and B.

Making $Cm = Cj$, $mg = jc$, etc., we will have mating points j and m, c and g, etc., in the peripheries of the non-circular wheels. That is, these points will come to contact in pairs, and in succession, on the line of centers as the non-circular wheels roll in mutual contact.

Following Fig. 106, we draw a normal Ca to the templet G, when a becomes one point in the required mating curve for B.

From j draw the normal jb to the templet G. This normal makes a certain angle with the radius Aj. Then draw mf, making the same angle with Bm produced, and make $mf = jb$. Then when the wheels revolve till j and m meet on the line of centers, the line mf will coincide with jb, and the point f with b. Therefore, f is a second point in the sliding curve for wheel B. Evidently g mates with c as a third point. Also h, i, and e are other points found in the same manner as was f; oi being equal to le, and the angle Boi being $180° - Ale$. Hence the sliding curve $iafgh$ may be drawn in, and a templet cut to the same, to be mounted upon the wheel or templet B. These sliding curves or templets thus made fast to their respective wheel templets A and B will evidently work harmoniously and agreeably throughout with the rolling curves or wheel templets A and B, and hence with a common law of angular velocity.

The extent of movement of these sliding curves is often limited, but a series of pairs may be employed which engage in action in succession, as in the case of the teeth of gear wheels, where the rolling curves tangent at C answer for *pitch lines*, and the sliding curves tangent at a for *tooth curves*. We therefore next take up the somewhat extensive subject of

THE TEETH OF GEAR WHEELS.

The sliding curves or tooth curves brought out above are sometimes called *odontoids*, from the Greek word for tooth. Also the rolling curves A and B in contact at C, the *pitch lines* in gearing, are sometimes called *centroids*, from the fact that lCj, etc., are points of instantaneous axes of rolling motion of the describing templet of Fig. 109 to be described.

THE TRACING OF SLIDING CURVES, OR ODONTOIDS, BY MEANS OF TEMPLETS ROLLED UPON NON-CIRCULAR WHEEL PITCH LINES OR CENTROIDS.

Templets are very useful aids in the study of or actual drawing of tooth curves for gearing, and we may employ pitch templets, tooth templets, and describing templets. Thus, a piece cut from thin wood to the form *AlCjck*, Fig. 108, may be called a pitch templet, and sometimes two pieces are prepared, one convex and the other concave, fitting each other along the pitch line *lCk*, either of which may be used at pleasure. Likewise, pitch templets may be prepared, fitting the pitch line *OCmgn*. Again, a piece, *G*, cut to the form of the tooth curve *eud*, and another for the mating tooth curve, may be called tooth templets. Besides these, a describing templet may be employed which, in Fig. 108, rolled along on the curve or templet *lCjc*, will, by an attached marking point, trace the tooth curve *eabc*, or again, rolled on *oCmg*, describe the mating curve *iafg*.

Having given the non-circular wheel pitch curve *lCjc*, to find the describing templet which, rolled along on the upper side of the pitch curve, will describe or generate the tooth curve *eabc*: Draw *aj*, *aC*, *al*, Fig. 109, of the same lengths as *bj*, *aC*, *el*, Fig. 108, and at intervening angles such that *lC*, *Cj*, etc., Fig. 109, shall equal the like lettered quantities of Fig. 108. Then draw in the curve *ajCl*, Fig. 109, and we have the figure of the describing templet sought, to complete which describing templet cut a piece of thin wood to the form of curve found and set a marking point at *a*.

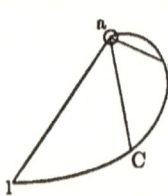

FIG. 109.

To trace the tooth curve *eabc*, Fig. 108, with this templet: Place the tracer, *a*, of the templet at *c*, Fig. 108, and proceed to turn the describing templet left-handed, causing it to roll along the pitch line *jCl* without slipping, when the points *j*, *C*, *l*, etc., of the describing templet, Fig. 109, will fall at the points *j*, *C*, *l*, etc., Fig. 108, by reason of the equality of spaces, step by step; and the tracer point *a* will at like points of time fall at *b*, *a*, *e*, etc., by reason of the equality of lengths of the lines *ja*, *Ca*, *la*, etc., Fig. 109, with the normals *jb*, *Ca*, *le*, etc., Fig. 108, and the equality of the angles in the two figures between arcs and lines at the points *jCl*, etc. This rolling of the describing templet may suppose that a concave pitch templet fitting the pitch line $cjCl$ is employed for the describing templet to roll upon, and the curve cae traced upon a plane which is fixed with reference to the pitch templet. In the illustrations, comparatively few points, c, j, C, l, etc., are indicated, in order to secure clearness of the figures.

For the same reasons that the tooth curve $cbae$ is traced by rolling the describing templet, Fig. 109, on the pitch line $cjCl$, likewise the mating tooth curve $gfai$ is traced by rolling the same describing templet on the pitch line $gmCo$.

It is to be observed that the above describing templet applies for only the portions of the tooth curves of Fig. 108 which lie above or to the left of the pitch lines. For the portions cd and gh, at the right or below, a second describing templet, found as was Fig. 109, would generally be required where, as in Fig. 108, one full tooth curve, e to d, was assumed. In practice, however, instead of assuming the tooth curve, the describing templet may advisably be assumed, and of any preferred form; and the same is often applied on both sides of each pitch line.

NAMES, TERMS, AND RULES FOR GEAR TEETH.

As the names of the various parts of toothed gearing will be of frequent use, they are here given, in connection with Fig. 110,

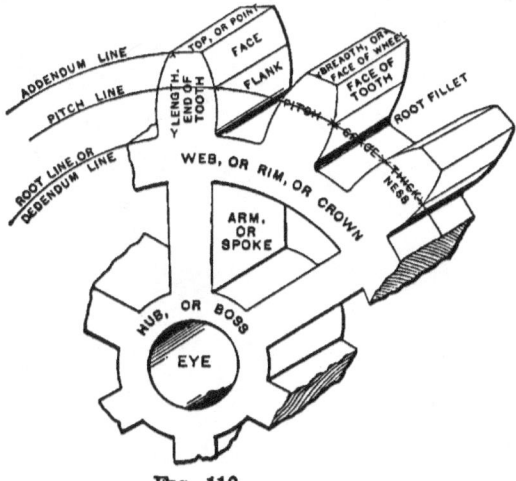

Fig. 110.

for future reference. It will be a help also in the study of teeth that the proportions be familiarized.

A common rule which may be deviated from according to the judgment of the designer is:

For Circular or Circumferential Pitch.

Pitch = distance from center to center of teeth on pitch line.
Addendum = 3/10 pitch; working depth = twice addendum.
Dedendum = 4/10 pitch, or subaddendum.
Thickness = 5/11 pitch.
Space = 6/11 pitch.

For Diametral Pitch.—For Cut Gears.

Pitch = portion of diameter to each tooth, = diameter divided by number of teeth in the wheel.
Addendum = one diametral pitch.
Dedendum = one diametral pitch plus 1/7 diametral pitch.
Thickness = nearly 1.57 diametral pitch.

CHAPTER IX.

TOOTHED GEARING IN GENERAL.

DIRECTIONAL RELATION CONSTANT. VELOCITY-RATIO VARYING.

TOOTH CURVES FOR NON-CIRCULAR GEARING.

Case I. Axes Parallel.

ALL correct working gearing of whatever name, circular or non-circular in outline of wheel, must possess tooth outlines which are describable by the principles indicated in Figures 108 and 109.

The most general case is taken up here, instead of several proceeding from the simpler to the more general, because the general case is as readily comprehended as any special limited one. Thus, the whole general theory of tooth outlines for gearing will be mastered at once.

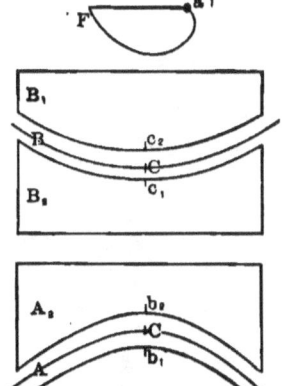

Fig. 111.

Templets.

The theory and practice for this case can probably be best explained by the aid of templets, a complete set of which, representing the general case of non-circular gears, is shown in Fig. 111. Portions of the non-circular rolling curves or pitch lines are shown as A and B; A_1 being a templet fitting the curve A on the inside, A_2 a second templet fitting it on the outside, both of which will fit together on the pitch line A. When thus together, mark the point C, b_1 and b_2 thus made coincident; also B_1 and B_2 are like fitting templets for the pitch line B, with points C, c_1, c_2 marked. These marked points are all mutually mating points as between templets or pitch lines. These are the pitch-line templets, and, for brevity, may be called pitch templets.

The smaller figures show the describing templets, aF and bG, with the marking point at a and b, and have their representatives, in theory, in Fig. 109. All these templets are supposed to be cut from thin pieces of wood, metal, or other suitable material.

USE OF TEMPLETS IN DESCRIBING TOOTH CURVES.

In Fig. 112, the describing templet aF is shown as being rolled along the pitch templet B_2, while the marking point, a, is describing the tooth curve c_1a; also the same describing templet is shown as being rolled along the pitch templet A_1 and describing the tooth curve b_1a. Now, if no slipping occurs in the rolling, and it be continued on both pitch lines till the arc c_1E be made equal to b_1D, then will D and E be mating points in the non-circular rolling arcs or pitch lines, and aE will equal aD both as to arc and chord, and the segmental figures aE and aD are equal in every respect. In this rolling, the tooth curves b_1a and c_1a are supposed to be traced by the marker a, and if extended, would give the dotted lines above a, a. These may be supposed to be traced upon the same paper that the pitch lines are drawn upon, and that the paper is transparent.

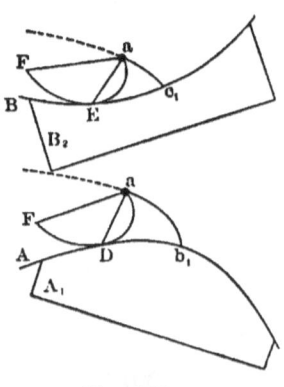

FIG. 112.

Now if we place one drawing upon the other so that they will be in common tangency with the points E and D coinciding, then the lines aD and aE will also coincide with the two points a in common; and the tooth curves b_1a and c_1a will touch each other at a. The resulting diagram for this will be as shown in Fig. 113; the two figures of the entire describing curves, indeed, coalescing into one at $aDEF$.

The two tooth curves b_1ad and c_1ae will be tangent to each other at a, instead of intersecting; because it is evident that if the describing templet $aDEF$ be rolled either way on the pitch lines, the point a will move to lower positions when the rolling is on Ab_1 than when on Bc_1. Also, it is plain that the line aDE is a common normal to the tooth curves b_1ad and c_1ae, since the point DE serves as a momentary center about which to describe small portions of the curves b_1d and c_1e at a, as in describing a circle about its center. These facts are confirmed by reference to Fig. 108.

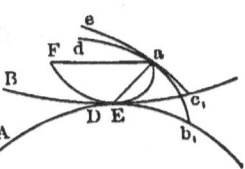

Fig. 113.

We observe further that the tooth curves b_1ad and c_1ae are wholly above their respective pitch lines and the point of contact, a, to the right of the line of centers through DE.

The principles brought out above being perfectly general, they embrace all possible forms of correct working teeth, such as epicycloidal, involute, conjugate, paraboloidal, octoidal, etc.

Hence all correct working tooth curves must possess the following properties:

1st. They must continually remain in mutual tangency.

2d. The mutual tangent point of a given pair of tooth curves must remain on the same side of the line of centers and of the pitch lines.

3d. The straight line joining the point of mutual tangency of the tooth curves and of the pitch lines must be a common normal to the tooth curves; and these normals should never intersect between the tooth curves and their pitch lines, and they must always intersect or be tangent to their pitch lines.

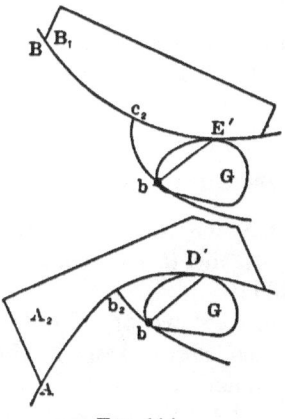

Fig. 114.

As Fig. 113 provides for contact only on the one side of the line of centers and above the pitch lines, contact below and to the left may be obtained by using the same, or any other, describing templet below and to the left, as in Fig. 114, where the other templet, bG, Fig. 111, is used, and rolled on A and B below as shown in Fig. 114 and combined as in Fig.

115, making the starting points, b_2 and c_2, coincide with $b_1 c_1$, Fig. 113.

This work, being all similar to that of Figures 112 and 113, will not require detailed description.

The Tooth Profile.

When combined with the points $b_1 b_2$, $c_1 c_2$, in common at C, at the line of centers, and omitting the describing templets, we have Fig. 116, in which LCM is a complete tooth profile for one side of a tooth of the wheel A, and NCO a tooth profile for one side of a tooth of the wheel B. Now, as the wheels turn one way from the position shown, the tooth contact moves

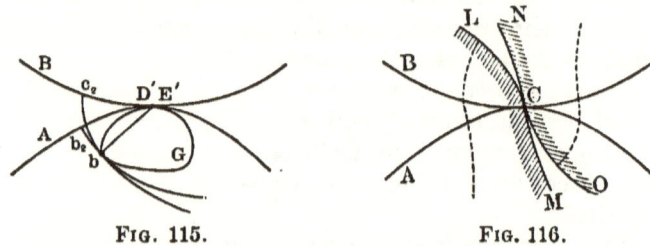

Fig. 115.   Fig. 116.

out in one direction from C, or the opposite direction for the opposite turning; from which it appears that in continuous repeated revolutions of the wheels A and B there will be contacts of the tooth curves as they approach the line of centers and also in recession therefrom.

These tooth curves in full lines in Fig. 116, correlated with other suitable ones in dotted lines, will furnish the outlines of a pair of teeth as indicated in Fig. 116.

POSITION OF TRACING POINT ON DESCRIBING CURVE.

The curves $b_1 ad$ and $c_1 ae$, Fig. 113, may be called trochoids when the curves A and B are not circular, the former being an epitrochoid and the other an hypotrochoid.

When the tracing point, a, is not on the arc of FD, but within or without, the curve generated is called respectively the prolate or curtate epitrochoid or hypotrochoid.

The tooth curves of Figures 113, 115, and 116 do not represent the most general case possible, since the points a and b are there carried on the curve, while it may be within or without. There-

TOOTHED GEARING IN GENERAL.   99

fore, let us examine the curtate and prolate curves in quest of advantages over those of Figures 113 to 116.

1st. *The Curtate Trochoidal Tooth Curve.*

Fig. 117 represents the case of the curtate curves, the tracer, a, being carried on an overreaching piece Ka, fastened to the describing curve FK, aK being normal to $KDEF$.

Take L and N as mating points in the pitch curves A and B, at which K is placed in starting the tooth curves, or trochoids, GIa and HJa. Then LG will be on a normal to A, and HN on a normal to B, and it is found that the portion HJ of HJa will interfere with more or less of the curve GIa, showing that the portion HJ cannot be used as a tooth curve. With HJ removed, GI may be also, as useless.

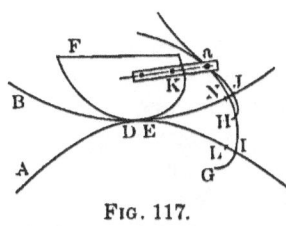

FIG. 117.

These curtate trochoidal curves from I and J upward will serve for tooth curves, and also others below A and B, but they will be found seriously inadmissible from the fact that there will be a wide space on either side of DE within which actual working tooth contacts are impossible, the very space where such contacts are by far the most desirable and valuable. The limit of this space to the right of DE is found by placing the tracer point, a, at J, and then noting the distance DJ. This will be found to increase with aK, and vanish only as aK disappears.

2d. *The Prolate Trochoidal Tooth Curve.*

In Fig. 118, the tracer, a, is placed on the normal Ka, then with L and N mating points at which K starts, as $KDEF$ rolls on A and B respectively, we will obtain the proposed tooth curves, Ga and Ha.

These curves will not interfere and may, theoretically, serve for teeth, but, practically, the curves are so nearly parallel to the pitch lines A and B that they will be more likely to slip and wedge the wheel apart with serious crowding, than to give desirable driving action. In other words, practical tooth curves should be perpendicular to the pitch lines rather than parallel to it, and hence the prolate trochoidal curves of Fig.

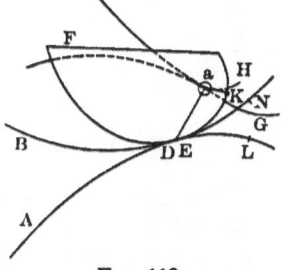

FIG. 118.

118 are inadmissible, though less so the shorter aK is made until it vanishes.

Hence the most favorable position of the tracing point or marker, a, is that it be exactly on the describing curve itself, as in Figures 113 and 115.

Form and Size of Describing Curve, or Templet.

Modification of the tooth curve, or odontoid, to a certain extent may be effected by employing one or another form of describing curve.

Thus, the describing curve may be a circle, ellipse, parabola, cissoid, spiral of many turns, or any arbitrarily assumed free-hand-drawn curve, each giving its own peculiar tooth curve when used on the pitch lines A and B; the flank curves apparently varying more than the face curves, as due to the changes in the describing curve as illustrated by the use of a small describing curve, Ia', or large one, Ja'', giving rise to the flank tooth curves Da' and Da'' respectively.

For gears engaging externally the faces will always be found as lying between the involute and pitch line, while the flanks may be anywhere from the pitch line on the one side to the pitch line on the other.

Thus, in Fig. 119, aD is the involute, described by rolling the straight line Fa on the pitch line A, the mating pitch line, B,

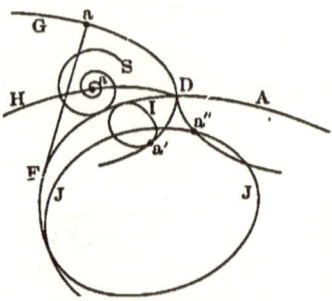

Fig. 119.

being in this case straight, as for a rack, in order that Fa may be used to develop the flank.

The curves above mentioned rolled along DF, starting the tooth curves at D, will evidently trace them all more or less below the involute DG, according to size, the smaller ones when re-entrant making one loop after another, according to number of convolutions on A.

These curves will all be normal to A at D except those developed by spirals which have an infinite number of coils around the marking point a, as in the case of the logarithmic spiral S with a at the pole. Thus the so-called involute teeth, with involute bases within A, have tooth curves which intersect A at an angle. Involute gear teeth, which at first seem to be an exception to the general theory of Figs. 113 and 115, are seen, nevertheless, to fall within the general theory, the tooth curves being generated with a log. spiral describing curve, rolled outside of A for faces, and inside for flanks, these flank curves being terminated exactly at the base circle, because here the radius of curvature of the log. spiral just equals that of the pitch circle A, upon which it rolls internally.

Example of Individually Constructed Teeth.

In non-circular gears, where the curvature of the pitch lines continually varies, no two tooth curves will be alike, so that every one must be described individually to be strictly correct. Fig. 120 is an example of a carefully made drawing of such gears, for

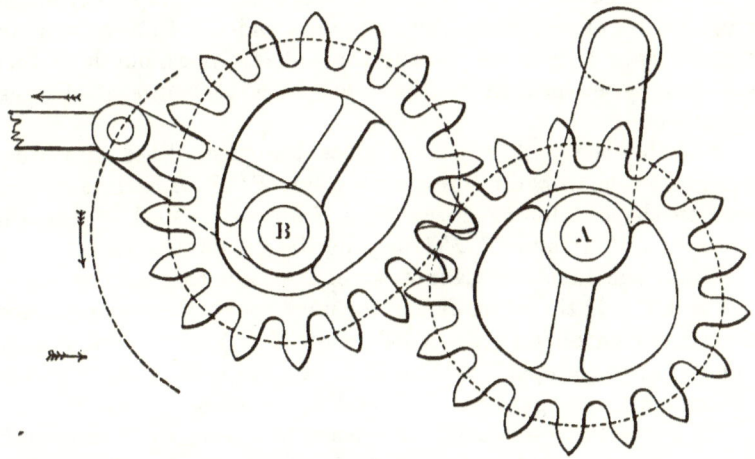

Fig. 120.

which the describing templets used for every face and flank were, for convenience, all circles, but varied in size according to the above principles, to favor the shapes of the teeth to prevent their being too weak across their flanks at points of sharpest curvature of pitch lines.

Non-circular Involute Gears.

In this case, as well as in circular involute gears, it is most convenient to make use of involute bases, or curves, a little within the pitch lines to which the tooth curves are involutes, as above suggested, instead of using rolling spirals.

Let Fig. 121 represent a pair of non-circular pitch lines to be set with involute teeth, and let the curves P, I, J inside of A, also R, L, M inside of B, be the bases of the involutes. The pitch lines are in contact at C, and for this position of wheels the bases of the involutes must be such that a common tangent PR will pass through the point C, and not only so for this one position, but for all positions of the wheels throughout the full extent of their rolling, whether segmental or otherwise.

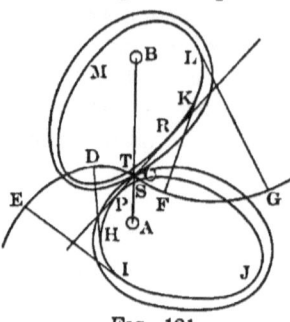

Fig. 121.

The inclination of the line PCR is arbitrary, except that undue irregularities of the base curves may-be avoided. In circular gearing the base curves are usually concentric circles, but here they will usually be nearest to the pitch lines at points of sharpest curvature.

Probably the best way is to place the pitch lines in various positions of mating tangency, draw a line PCR for each, and then trace in the enveloping curves P, I, J for each wheel. Then the involutes $SCDE$ and $TCFG$ may be drawn for the teeth. It is seen that the involutes cannot extend within the base curves as at S and T. If the addenda call for more room, the space between teeth may be arbitrarily deepened below S or T to a suitable root-line depth; but proper tooth contacts cannot go below these points, nor beyond the tangent points P and R, by reason of interference.

These teeth are less advantageous for gears of considerable eccentricity than those of epitrochoidal form of Fig. 120, because of the greater crowding action explained in Fig. 128, but they have the advantage of requiring less attention for maintaining exactly the distance A B between centers.

Conjugate Teeth.

This practical method of constructing tooth curves where one of a pair is arbitrarily assumed, and the other determined from it, would seem to contradict the refinements of the above theory to the extent that one tooth may be compelled to take any preferred form under sheer caprice.

Thus, prepare the pitch templets A and B, Fig. 122, and fasten to B a sheet of paper D, and to A a tooth templet EF elevated sufficiently above A to allow the paper D to go in between A and F.

Then bring the two templets A and B into proper contact at C, Fig. 123, and while held securely draw the pencil along the tooth-

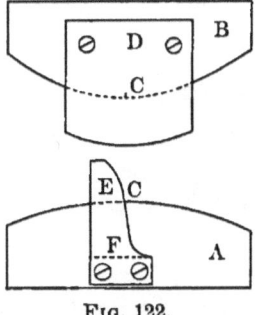

FIG. 122.

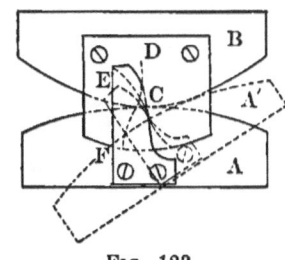

FIG. 123.

curve edge of the tooth templet ECF, copying its shape upon the paper D. Then roll A upon B some distance without slipping, as indicated by the dotted line at A', and while held fast, draw another line along the edge of ECF. Thus draw several lines for A in positions to the right and left, when D will present a series of lines, shown dotted in Fig. 124. Now, the enveloping curve GCH, or full line just touching all the dotted lines on D, will be the best mating tooth curve for ECF. All the tooth curves for one complete wheel might be thus assumed, and the mating ones for the other wheel found in the same way.

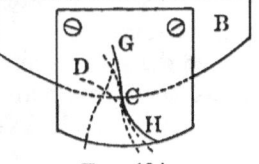

FIG. 124.

This process is seen, after all, to present nothing but what is disclosed in Fig. 108, since G there corresponds with F in Fig. 123, so that a similar process carried out in Fig. 108 would result in the curve $hgfai$.

Limited Inclination of Tooth Curves.

In the practice of designing non-circular gears of considerable eccentricity, it will be found very desirable at times to make the teeth nearly radial and oblique to the pitch curves, as in the full lines of Fig. 125, instead of normal to the pitch curves and inclined to the radius; thus to prevent the tendency of the teeth to slip out of engagement, as well as the excessive crowding apart of the gears.

Therefore, let Fig. 125 represent portions of a pair of rather eccentric non-circular pitch curves A and B, with a proposed pair of teeth E and G in contact at the line of centers C, which, in the

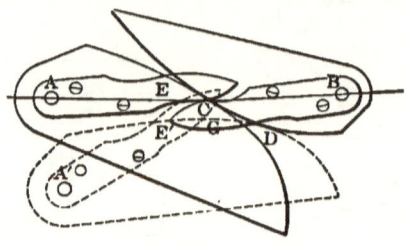

FIG. 125.

light of assumed tooth curves of Fig. 122, and in spite of the theory previously detailed, are set nearly radial to avoid crowding.

Now, with B stationary, roll A around, without slipping on B to the dotted position A', with E following around to E'', shifting the contact of the pitch curves from C to D. Then it will be found that the tooth E at E' fatally interferes with the tooth G, showing at once that the amount of departure from the pitch curve normal toward the wheel radii shown in the figure is absolutely prohibited.

In some actual wheel designs where this partiality for the radius existed, the teeth necessarily were doctored before the wheels would work.

Thus, in Fig. 126, we have a photo-process copy of a pair of gears made for practical use in machines, examples of which were

FIG. 126.

exhibited at the Centennial of '76, in which an effort at inclining the tooth toward the radius, giving it a hooklike form, was evidently attempted, but was vetoed by the theory when the wheels were put to work, as is plain by the fact that the hook feature was removed from the hook or overhanging side. One interesting fact incidentally connected with these wheels is that they are cut gears, the gear cutter to cut them having twenty-five cutters. Thus we have a rare example of other cut non-circular gears than elliptic.

Other examples of the endeavor to obtain hooklike teeth were found at the Centennial; one showing this peculiarity in a moderate degree being illustrated in Fig. 127.

Another example was found in a shingle-sawing machine from Wisconsin.

In all the above discussions, from Fig. 92 on, we have treated the case as if the wheels were plane figures without thickness, while in reality we may suppose them to have any arbitrary thickness as cylinders, transverse sections of which form the above figures.

FIG. 127.

Nearest Approach of Tooth Curve toward the Radius.

To determine absolutely the nearest approach of the tooth curve to the radius, we have only to refer to the principles laid down as following Fig. 113, and especially that of the intersection of the pitch line by all the normals to the proposed tooth curve. Thus the curve $c_2 b$, Fig. 114, will become more and more erect, with respect to the pitch line BE', the farther the intersection E' is kept away from the point c_2: and this is favored by a large curve bG, the largest being the straight line. But this same curve must roll inside of A, and hence the curve A itself is the largest possible one. For this, however, the flank $b_2 b$ would be merely a point, and hence objectionable from considerations of wear.

It therefore appears that a sharper curve than A is desirable.

In Fig. 128, A and B is a pair of non-circular wheels where the attempt is to be made to carry the tooth curve GC as far as practicable over toward the direction of the radius BC, changing GC to some position NC as dotted.

First, drawing a normal NO to the proposed dotted curve CN, this normal is found to escape the pitch curve CL altogether, and hence the curve CN is steeper than admissible, since the normal must intersect the pitch curve CL.

Next, we propose a describing curve, E, of large radius, such as will but just go within the mating pitch curve CM and roll, giving the shortest admissible flank CI.

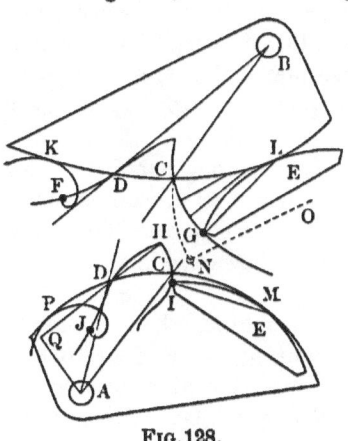

FIG. 128.

This describing curve E must be accepted as that which will give the greatest permissible deviation of GC toward NC.

This tooth curve GC is, however, far from coinciding with the direction of the radius BC, and hence as the flank CI drives against the face GC, there will be a more or less serious crowding apart of the wheels A and B.

Wear in practical use of wheels is the criterion by which CI is limited in length, it being, at best, much shorter than CG against which it works. Good judgment would protest against CI being reduced to a point, as above suggested, for the case where E conforms with CM; and also the figure shows that this only very slightly moves the curve CG from its present position toward CN.

Hence, though the utmost possible careening of CG toward CN occurs when CI is reduced to a point, yet practical considerations favor a slightly less elevated curve CG, to prevent an appreciable length of flank CI; and we conclude that a tooth curve CG, lying near the radius BC prolonged, is impossible.

On the other hand, the tooth curve may be inclined the opposite way, or toward the pitch line, to any desired extent, as shown in Fig. 128 at DH and DF; it being simply necessary to make the describing curve, for example $FK = JP$, quite sharp in curvature, or small, and a spiral of several turns about the pole J for inclination of tooth-curve at D.

Practical Limit in Eccentricity for Non-circular Gears.

In theory there is no limit short of a radius for each wheel; but in practice friction interposes a limit, the full investigation of which is out of place here, so that a few examples, only, will be cited to indicate how essential are the principles of Fig. 128 in gear designing.

Fig. 129 shows a pair of gears made for use on a machine to run at about 350 revolutions per minute. They are $4\frac{1}{4}$ inches between centers and have involute teeth, both being cast in iron from one pattern and to work without lubrication. They were laid out after the principles of Fig. 79.

When started in the newly developed machine, whole segments of the rims of these gears would drop out upon the floor in less than two minutes, though believed, as far as ordinary theoretic strains were concerned, to be ample in strength. On examination, it was found that the involute tooth curves were too much inclined to the pitch lines, so as to actually block the wheels at positions where the radius was also much inclined to a perpendicular to the rim.

FIG. 129.

It was found that elliptic wheels would give a more favorable obliquity of pitch line to radii, and that teeth generated by large describing curves like E, Fig. 128, would give easier action; whereupon new wheels were adopted with these changes and with coarser teeth, a photo-process copy of a pair of which is given in Fig. 130.

FIG. 130.

In these wheels every effort was made to provide against breakage by "blocking," two considerations being included besides those above mentioned, consisting, first, in thicker teeth on the sides of the wheels than at the ends, and, second, shorter teeth, compared with pitch, than for usual proportions. The teeth are also much coarser than in the previous wheels of Fig. 128, so much so that but little room was left for arms and openings.

A quite notable case of very eccentric gears is given in Fig. 131, made in cast iron, 2 feet between centers, the larger being some $3\frac{1}{2}$ feet long, and used on a hay-press, where the larger wheel is driver. Here the teeth are of unequal size as well as in Fig. 130, the proximity of the axis in the small wheel controlling the size at the end. The tendency to block in action is apparently great, as due to the unusual length of the teeth, they being here extended to sharp points.

An examination of the figure indicates that for the large wheel as driver the blocking tendency at places is very great, so much so that thorough lubrication must be a necessity to bring the coefficient of friction down to permit the wheels to work at all.

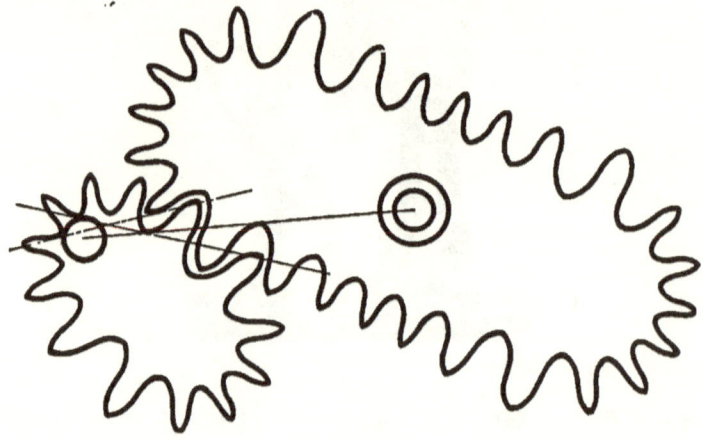

FIG. 131.

Fortunately for this case the wheels are used in a horse-power machine, where, if the wheels block from deficient lubrication, the horses may stop as a sign for lubricant wanted, the wheel having sufficient strength to resist the horses. In a machine with the inertia of a heavy fly-wheel or other parts involved, the result would be different.

The full line drawn as a common normal to a pair of teeth just taking contact, with the large wheel driving right-handed, strikes very near the center of the driven wheel, making a case of positive block with coefficient of friction at 0.083. As the coefficient of friction is about 0.12 to 0.16 for dry cast iron, and 0.05 to 0.08 for lubricated, it is clear that these gears cannot run except with teeth well lubricated.

In the case of the dotted line for the next tooth there is still greater doubt.

If these teeth were cut off to the ordinary addendum of 3/10 pitch, the tendency to block would be very much less, and still better if cut to the minimum where one pair of teeth engage in action as the preceding pair quits engagement, and the wheels would then probably run without lubrication.

Whence it appears that unduly long teeth, such, for example, as run to a point, are especially objectionable in very eccentric non-circular gears.

Again, where one of a pair is always to be driver, the addendum of the driven wheel may with advantage be made unduly short with that of the driving wheel increased.

In heavy gearing rigidly connected with heavy revolving parts, the ratio AQ over QH, Fig. 128, where QH is a normal to CH at H, should not exceed 0.4 to 0.5 for teeth running without lubrication.

It therefore appears that there is a practical limit of eccentricity in non-circular gearing, occasioned by the "blocking" tendency at points where the pitch line departs farthest from a normal to the radius, and that this limit is favored by short addendum of teeth of driver; by large tooth-generating curve for side of tooth in question; by thick and stout teeth; by lubrication; and by giving the driver a slightly greater pitch than follower.

Substitution of Pitch-line Rolling for Teeth in Extremely Eccentric Gears.

In some cases of very eccentric gears, where blocking would surely occur with teeth, the latter may be omitted with advantage and the driving action from driver to follower effected by the mutual rolling of the pitch lines.

Fig. 132

Fig. 132 is an example of this, which represents the limiting case of the "quick return" movement, where a crank and pitman is made to drive the connected slide forward at a uniform velocity and back in much less time, or "quick." The model of Fig. 132 represents the theoretical limit of uniform motion forward from beginning to end, and back in an absolute instant, or no time at all. This is evidently impossible, since the driven wheel must make a half-turn in the absolute instant, during which the driver would be stationary with no power to drive.

In the model this difficulty is removed by placing the handle on the driven-wheel crank or lower wheel in the model.

In the absence of teeth, the eccentric pitch surface of the driver acting upon that of the follower can drive in one direction, but not in the other, and to prevent derangement some device must be supplied.

In elliptic gears a link may be added as in Fig. 133, where the eccentricity is too great for good working teeth throughout, so that a pair, only, of teeth is added at the ends and the link introduced to hold the elliptic pitch lines in rolling contact.

But in Fig. 132 a link cannot be employed, so that a circular sector with non-circular arcs at the ends is added to engage with a mating piece on the opposite wheel. These are so shaped as to hold the upper wheel stationary for the allotted time of the half-turn of the lower wheel, after which the non-circular curves keep the rolling pitch-line arcs in contact for the space where teeth are omitted.

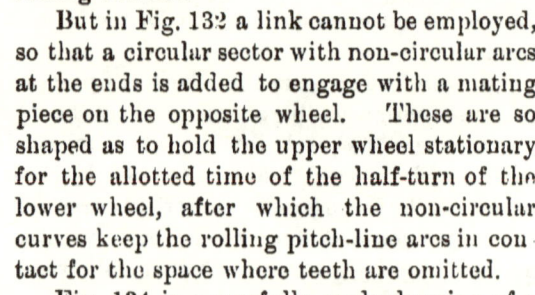

Fig. 133.

Fig. 134 is a carefully made drawing of a pair of gears used for a while in a certain screw machine for a "quick return" to dirve a slide uniformly forward and back with the least allowable time greater than an absolute instant. It is produced, in effect, by modifying Fig. 132, as by splitting the long projecting point of the upper wheel and spreading it to DAE by a circular sector, mated with FCG in the other wheel, the two wheels being otherwise modified in a manner to preserve the quick-return principle exactly. As adopted, the advance occupies 3/4, and the return 1/4 of the time. The line of centers is about five inches.

From H to F, and G to I, the pitch lines roll upon the mating

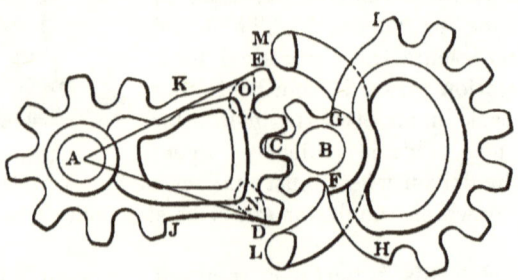

Fig. 134.

pitch lines DJ and EK, respectively; teeth engaging for the remaining arcs.

To prevent the gears from getting deranged in this case, projections from A, shown by dotted lines at N and O, are added, which fall inside of projections *L* and *M*, respectively, and serve to keep the pitch lines approximately in contact, and assuring the engaging and disengaging of teeth for several hundred revolutions per minute in either direction.

FIG. 135.   FIG. 136.

Fig. 135 gives a better view of the projections *L*, *M*, *N*, and *O* of Fig. 134.

These wheels served with entire satisfaction, as operating gears, for a considerable time in the machine they were designed for. But as the return, though decidedly more moderate than in Fig. 132, was yet decidedly too quick for the purpose, other wheels with still less rapid return were found advisable, and obtained in the wheels shown in Fig. 136.

Here the return was so extended that one of the rolling arcs was dropped, the other being necessarily retained to preserve the law of uniform motion of the slide, the one set of projections *L* and *N* being also retained to provide against misengagement of teeth.

These wheels, as in Fig. 136, are doing good service in extended use and are thoroughly practical gears.

Internal Non-circular Gears.

These are practical for complete wheels where the "internal" gear or that having teeth within the rim is at least twice as large as the pinion, or smaller wheel; and the tracing of the teeth is as before except for the fact that

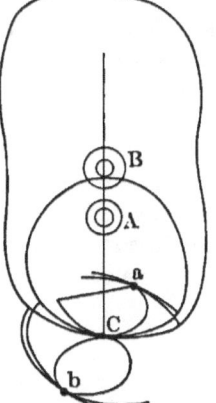

FIG. 137.

generally both pitch lines are convex in the same direction as in Fig. 137, so that a further description is unnecessary.

CHAPTER X.

TEETH OF BEVEL AND SKEW BEVEL NON-CIRCULAR WHEELS.

CASE II. THE GENERAL CASE OF AXES INTERSECTING.

WE here assume that the pitch lines are already made out actually on the *normal spheres*, or else as ordinates, angles, etc., by which the curves may be drawn, as explained under Case II of *Rolling Contact, with Axes Meeting.*

Let Fig. 138. represent a pair of such rolling wheels with centers at A and B, and with C the common tangent point. Such

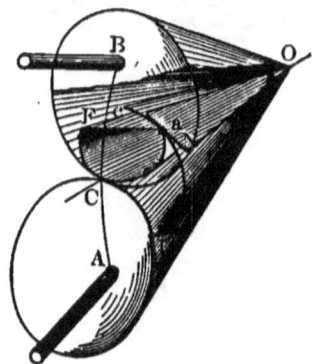

FIG. 138.

wheels with axes meeting are conical as explained under rolling contact, and may extend in thickness to the vertex O of the cones. The front surfaces of the wheel at AC and CB are spherical surfaces, O being the center of the sphere and the common vertex of the cones. The finished wheels are usually as if cut from spherical shells as in Fig. 83.

The describing curve or templet to carry the tracing point is here a cone, also, with base aCF, and with vertex at O in common with the other cones. The tracer or marker is here a line aO.

Let L and N represent mating points on the rolling cone-wheel bases, LC and NC being equal. Then placing the line aO of the describing cone coincident with the element LO of the wheel cone, and rolling it around to the position aCF, we describe a surface $LaOL$, which may be extended to c as shown. Also placing the same describing cone inside the wheel B, and with the line aO coincident with the element NO of the wheel B and rolling it around to the position aCF, we generate the surface $NaON$, which may also be extended by further rolling. These two generated surfaces will be tangent to each other along the element aO, and hence these surfaces form proper tooth surfaces for teeth of these wheels, whatever their thickness.

In the case of spherical shells, the tooth curves may be traced directly on them, both sides if desired, by means of rolling templets of wood cut concave to the spherical form and of any preferred outline aCF.

In Fig. 138 we have a face for a tooth of A, and a flank for a tooth of B, contacts between which will always be inside of B and to the right of ACB. Contacts on the other side may be obtained by rolling the templet aCF, or any other one, inside of A and outside of B, thus procuring a flank for A and a face for B.

For non-circular wheels with large teeth it will be necessary for accuracy to trace every individual tooth curve for the entire wheel, no two curves being alike; but the same templet will not be required except for the pair of tooth curves that are to work together, as La and Na in Fig. 138.

In small teeth less care is needed and approximate tooth lines will serve, as, for instance, for 1 or $1\frac{1}{2}$ inches pitch or less. In some cases, as for instance in Fig. 98, where the gears were cast, the patterns' had the rim formed to the root line, when teeth blocked out for the full depth in separate pieces, by a drawing giving some excess of curvature at the sides, were fastened upon the rim and the patterns finished and the gears cast.

The Tredgold's approximate method, hereafter described fully in connection with circular bevel gears, may be employed here with sufficient accuracy for any practical purpose.

Involute teeth may be laid out by first finding the base lines and using rolling planes on the base lines, similarly as lines are used in Fig. 121, the tooth surfaces being generated by a line on the rolling plane running to O of Fig. 138.

Case III. The General Case of Axes Crossing Without Intersecting.

NON-CIRCULAR SKEW BEVELS.

A method for the exact construction of teeth for this case is not known, nor even for circular skew bevels for serviceable gears, except such as are taken from near the shortest common perpendicular between the axes.

Suppose the pitch curves have been determined upon the spherical blank as at p, Fig. 139, by Figures 93 to 95. Then the addendum and dedendum lines may be laid out as at d and e, Fig. 139. Between the last-named lines the tooth curves are to be drawn as shown by the method of Fig. 138, rather small describing curves being used to prevent the teeth from interfering unduly by reason of the admitted approximate form of these teeth for non-circular skew bevel.

Then, as explained in connection with Fig. 95, construct the second spherical blank M to mount upon an axial rod equidistant with it from the gorge, the teeth being laid out upon M in the inverse order with respect to A; or in like order as the blanks are viewed in Fig. 139, the convex side of A and the concave side of M being towards the eye.

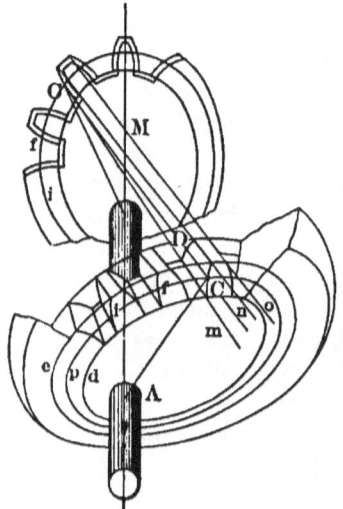

Fig. 139.

Cut the teeth on M to line and bevel back towards the concave side so that the line or string CC, drawn tight, will spring from the line of the tooth curve on the convex surface of M.

Then the blank A may be cut to the addendum line and to the tooth curves by stretching the string across, at various corresponding points of A and M, as at A, m, n, o, etc., and at as many intermediate pairs of points as desired.

These teeth will be twisted, but the face and flank surfaces will be composed of straight-lined elements.

Instead of cutting these teeth from the solid material of the blank A, the latter may be dressed off carefully to the dedendum line and separate blocks for teeth may be prepared and fastened upon the blank, and finished subsequently unless quite small and short along CD.

The wheel B, mating with A, may be made in a similar manner. If they are of the same size and alike, one pattern for casting will serve for both A and B, but otherwise not. The wheels may, either of them, be made 1-lobed, 2-lobed, etc.

When a pair of wheels are thus far completed they will not run together smoothly because of the already-mentioned incorrectness of the tooth outlines for this case.

The running conditions may be improved by mounting the wheels as mated in mesh, and by revolving slowly under careful scrutiny, the high or interfering points noted, and dressed off by hand, this process being continued till the desired degree of smoothness of running is obtained.

The resulting wheels, obtained as above, may be of wood and used for patterns for casting in metal with no greater further expense than for circular wheels.

For wheels of one inch pitch or less it is probable that teeth may be formed in separate pieces and made fast upon the blanks A and B, and dressed to the dedendum line and surface.

CHAPTER XI.

INTERMITTENT MOTIONS.

TEETH; TOGETHER WITH ENGAGING AND DISENGAGING SPURS AND SEGMENTS IN GENERAL.

TEETH.

WE now come to consider the teeth and spurs of intermittent motions whether having circular or non-circular pitch lines.

The teeth present no new problems over segmental or partial gears, as, for instance, in Fig. 96, the pitch line FED when equipped with teeth is as a segmental gear, to mesh or mate with the segmental gear GMH, and any preferred form of tooth may be adopted and laid out as already explained. There should, however, be a space to start with at F and D and a tooth at G and H, as here shown in Fig. 140 at N and K.

Directional Relation Constant.

THE ENGAGING AND DISENGAGING SPURS.

The spurs GF and $H1$ are to start the driven member B from the full stop on the locking arc into full motion, as for the initial engagement of the teeth, and this usually by contact action of the spurs. Where a sufficiently large arc of movement can be allowed to A in which B is to be started or stopped, the spurs may work by rolling contact instead of sliding.

In Fig. 140 we have a correct drawing of a pair of non-circular plane wheels of the intermittent order, showing two complete sets of prongs or spurs for engaging and disengaging. One set, numbered 1, works by sliding contact both ways, and the second set, numbered 2, works by rolling contact; the same arc of movement ab for wheel A being required to swing B from rest into full engagement when the sliding-contact spurs are employed as when the rolling contact spurs 2 are in use. The same is true for the spurs on either side of the wheel, the arc cd being greater than ab, as required by the greater drop MN in radial distance.

The spur *HI* mates with *F'G*, and they are so made that when the latter approaches the former, *F'* and *H* are the first points to touch; this same contact starting *B* towards engagement of the gear teeth, when further movement causes the point of contact to move along toward *G* and *I*, arriving at which the full engagement of the gear teeth should be assured.

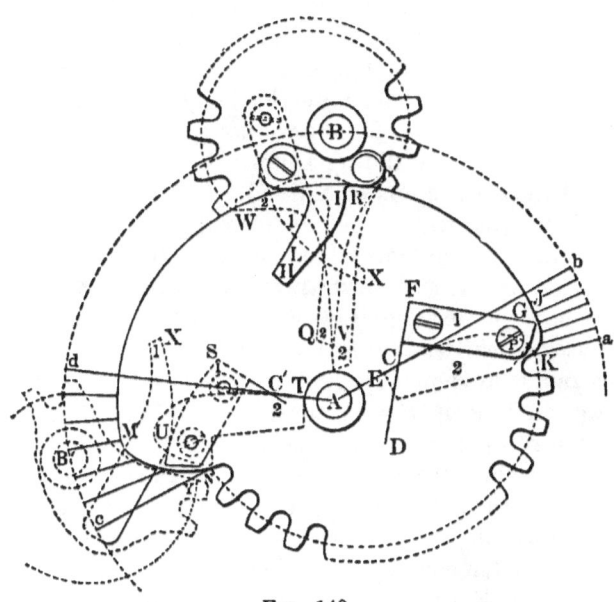

FIG. 140.

Where this movement takes place the circular locking arc *JM* must be cut away as at *JK*, and to just such extent that the wheel *B* shall not have undue backlash, it being prevented by the bearing of *B* upon *JK* for the one direction, and by the contact between *FG* and *HI* for the other direction. The same, of course, is true of the other side of the wheel *A*, where *B'* is shown as bearing against *MN* for the one direction of motion of *B'* and against the spurs *ll* for the other direction. It is essential for the satisfactory performance of these wheels that these curves and spurs be well mated with respect to reduced backlash.

For rapid rotation the spurs need to be much the longest. For slow motion a pin may do for *FG*, with *HI* cut shorter to match, as shown at *D* and *F*, Fig. 141, where the shock is some 4 to 6 times as great as in Fig. 140. But the shock is materially reduced by giving the point of the spur a curve as shown in the example at E and H, Fig. 141, the intensity of the shock being judged of by the perpendicular distance from A to the normals shown. Some shock will occur at all events when the points H and F, Fig. 140, meet in contact unless HI and FG are so much extended that the normal FD to GF meets the center A. To determine the amount of shock in a given case, we are to consider the segments AC and bC of the line of centers Ab at the initial contact of H upon F, these segments being formed by the common normal FD.

In EP and QR, Fig. 140, we have a pair of spurs working by rolling contact instead of sliding. These differ from the sliding spurs in being much longer for a like tendency to shock, and consequently objectionable, though the rolling action is in their favor in respect to wear. To determine the lengths for equal shock we note that the distance AE for the rolling spurs should be the same as the length AC for the sliding spurs. The same is true for the terminals of the other spurs at SX and VT.

The rolling spur QR, which mates with EPK, is unavoidably concave because the contacts which occur on QR must be progressive from Q toward R, and must always be on the line of centers for rolling action; and for a gradual start of B from rest, the line QR near Q must be nearly radial and recline more and more with respect to a radius as shown by the line QR, and, in fact, should run into the pitch line of B as a continuous rolling line. Likewise EPK should run into the pitch line KN.

Again, that the acceleration be normal, the curve EP should gradually turn over into the pitch line for the teeth at K, as if it was the termination of that line toward A, both portions EP and KN being, in fact, a continuous rolling arc, and, reasonably enough, should begin at E near A, and by continuous smooth curvature pass the points P, K, and on to N. Thus this half of these wheels resembles the half of the wheels of Figs. 132 and 134, where a pair of rolling curves run into the pitch lines of gear teeth.

At MN the locking arc is necessarily flexed to a greater distance toward A, by reason of the non-circular pitch line NK employed, and MN is shaped as due to an effort to give B a quick move from lock to engagement, as may sometimes be necessary in the application made of these wheels. To this end, it is here found necessary to give B a more rapid motion for a part of this turn from rest to engagement, than it will have while engaged with the teeth. This is not so evident from the sliding spurs S and X as from the rolling ones TUN and VW, for which the law of angular velocity is made the same. Examining the latter, we note that for contact of VW on TU from T up to U the acceleration is rapid, and that at U the speed of B is full double that due to contact at N when the gear teeth have fairly engaged. One effect of this is to leave the tooth Y behind with liability of its interfering with the tooth it is approaching. Again, it appears prejudicial to accelerate the motion of B to 100 per cent. in excess and retard it again, the only excuse for which in practice is probably to be found in a necessity for a quick movement of B from rest to the full engagement of the gear teeth, or for a comparatively small arc cd for it.

At all events it is certain that the rolling spurs TUN and VW are utterly impracticable at least from U to N, where the driving action of the spurs would be negative.

That rolling spurs may be made practical in place of TUN and its mate, Fig. 140, we infer from the curve EP, that instead of going outward at first to U and then turning in toward A to N, it should start at once from T on a curvature turning toward N. But an examination of the case will readily show that this is impossible unless the arc cd is increased.

In Fig. 141 the problem of practicable rolling spurs for wheels A and B is undertaken, where NBV is the total angle through which B must turn in going from rest to engagement, and NAd' the corresponding angle for A.

The angle NBV is divided into a series of decreasing angles from N toward V, then the same number of angles are laid off from NA toward Ad, all of which span slightly more arc on the radius AN than does the first of the series NV on the radius BN. Then the curves are drawn in according to the method of Fig. 69, giving the rolling spurs NUT and NGV.

It is plain that the portions of the rolling curves NG and NU are so nearly perpendicular to the radius as to be likely to work bad in practice unless provided with teeth, becoming part of the toothed segment, and that the easement curve is very much prolonged as compared with that of Fig. 140. Also it seems that the speed of B in some cases must be accelerated beyond that of the gear teeth in engagement and then retarded to the latter velocity.

The arc of engagement $e'd'$ is here very wide, as due first to the very considerable drop in the radius from AH to AN, and, second, to the fact that AN is so short compared with AB.

It appears, then, that rolling arcs may always be adopted for engaging or disengaging spurs of intermittent wheels in place of sliding arcs for the same; but that the former are subject to some prejudicial features not inseparable from the latter, but which outweigh the objection to sliding contact action of the latter.

These wheels may be made as bevels for the case of axes meeting, a complete example in metal being shown in Fig. 98, which worked satisfactorily.

In skew-bevels they are doubtless possible, but probably after a tedious time in construction they would be found unsatisfactory mainly by reason of the sliding along the line of contact of the various surfaces, requiring the spurs to be very thick, etc.

In Fig. 140 the wheel A is shown with only one locking arc, and B one locking segment. But either may have two or more, involving, however, no new problems over these already considered, there being a toothed arc followed by a locking arc, and the latter again by a toothed arc, etc., in succession, spurs being provided throughout.

In case of slow movements of this kind the piece $F'G$ or S in Fig. 140 may be made shorter, indeed reduced to a mere pin as at E and H, or F' and D, Fig. 141, thus simplify-

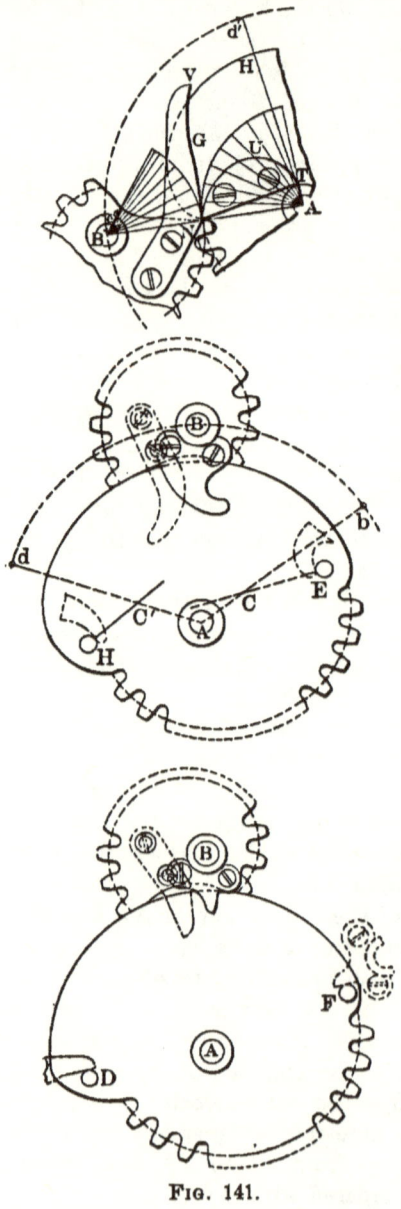

Fig. 141.

ing the movement in construction.

By giving the spurs a very considerable curvature, as at E or a H, the shock due to contact of spur and pin will be materially reduced as compared with that attendant upon the spurs and pins at F and D. Here the normal EC or HC' is to be drawn, and the segments AC and bC measured by which to judge of the shock at contact of pin and spur. The shock will depend upon the velocity-ratio at the instant of initial contact, that ratio being $\frac{AC}{bC}$ or $\frac{AC'}{dC'}$. Comparing with Fig. 140, it seems possible that pins and spurs may be so made that the shocks are no more prejudicial than in Fig. 140.

SOLID ENGAGING AND DISENGAGING SEGMENTS OF INTERMITTENT MOTIONS.

Intermittent motions with spurs attached are objectionable in coarse and heavy machinery from the tendency to loosen and become detached. These wheels, of solid casting complete, are much the more satisfactory in such places as binders for reaping machines, etc., both for simplicity and cost, especially where they are small or run slowly. By suitable shapes of the engaging and disengaging segments, however, they may run at speeds equalling those provided with spurs, as in Figures 140 and 141. (See Figures 15 and 16.)

A suggestion as to the shapes of the locking segment of B is obtained from Fig. 141 at E and H, where the spurs might seemingly be in the plane of the wheel B, and A correspondingly cut away to a circle arc coinciding with the prongs of B as proposed, the latter so connected as to form the locking segment for B. This would lead us at once to Fig. 15. Following the suggestion, we obtain Fig. 142 as a pair of these wheels of the non-circular form, B being at rest on the locking arc.

In designing these wheels, the distance IL and LK should be sufficient to prevent the tendency of the ends of the prongs I and K from biting into the arc DLE and blocking the movement, or, in case of heavy wheels, causing breakage. To examine into this, we will find that the angle BKM must exceed the so-called "angle of repose" for the material and conditions at K. A portion of the arc at J should be cut away to prevent "clinging" to the arc IK upon the terminal points D and E of the arc DLE.

Between EG and DF a portion is cut away to allow the prongs of B to pass as the latter is started into motion. These gaps are undercut at G and F to reduce shock due to initial contact with the prongs of B, the intensity of which shock may, as before, be examined relative to the distance AP or AQ from the center of motion of A to the normals at F and G. This shock is zero when AP and AQ are zero.

No direct and simple rule of graphics is known for laying out the wheels of Fig. 142, beyond that for the toothed pitch lines,

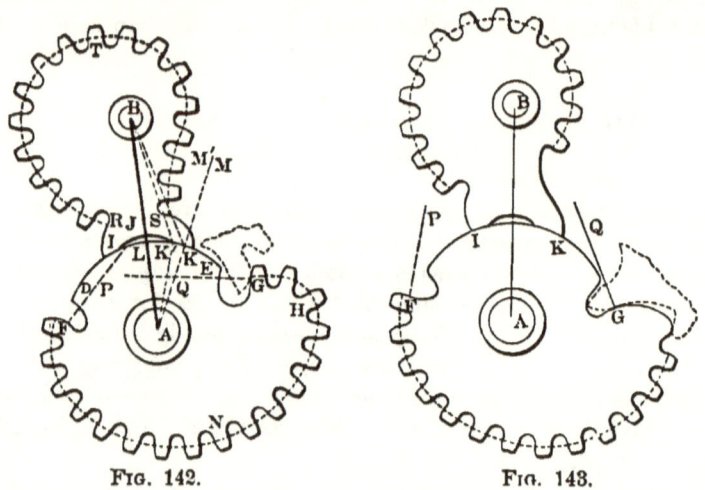

FIG. 142. FIG. 143.

where the arc FNG must equal RTS. The radius AL is arbitrary, except that it ought to be less than either terminal portion of the toothed arc $FNHG$. After assuming a trial arc DLE, and with trial curves for the wheel A at F and G sketched in, a circle may be drawn through B as at rack for the latter and a trial templet in cardboard cut out by guess for BIK, with a hole having a point at B for center. This templet with guess forms at IR and KS is to be tried at several positions as shown at EGH, the center always remaining on the circle arc through B. Likewise at FD. A guess shape of the prong of B at I may be tried, and if found unsuitable a modification suggested by this trial may next be cut out and tried. By this tentative process the most suitable shapes at IR and KS and at DF and EGH are to be determined. The curves FHG and RTS must be correct rolling arcs for gear teeth. Several trials will probably be needed to bring out the most satisfactory shapes at F, I, K and G.

For slow movements, where the shock of starting and stopping of B due to speed is at a minimum, the prongs I and K may be less hooking and simpler, as in Fig. 143, for which, at a given speed, the shock is about four times as great as for Fig. 142.

The intensity of shock may be examined into here by drawing the normals FP and GQ and considering the lengths of perpendiculars from A upon these normals. These are seen to be greater in Fig. 143 than in Fig. 142 and hence the greater shocks.

Bevel wheels after either variety, Fig. 142 or Fig. 143, may be readily worked out according to Figs. 85 and 87 after the plain wheels are drawn, as illustrated by Fig. 98 of a working model in brass. Skew bevels are here much more practicable than in the wheels of Figs. 140 and 141, and may be worked out according to Figs. 93 to 95, and be practical in running machinery.

FIG. 144.

COUNTING WHEELS.

These wheels, of high durability, may be constructed as in Fig. 144, which is a positive movement with no possible likelihood of derangement except by breakage. For counting, the upper wheel should have 10 teeth. The "Geneva stop" is a familiar example of this class of wheels.

ALTERNATE MOTIONS.

Directional Relation Changing.

The pitch lines for these wheels have been treated in Figs. 99 to 103.

The most essential features for successful wheels of this kind consist in the peculiar shapes of terminal teeth, or of segments, or of spurs, or of shifting devices, as in mangle wheels.

These wheels may be divided into two distinct classes, viz., as those admitting of limited movements only, and those of indefinite extent of movement, to be taken up in succession.

FIRST. LIMITED ALTERNATE MOTIONS.

The movement is in this case limited because the driven piece must unavoidably make a forward and return movement for each revolution of the driver. The pitch lines of these wheels have already been considered in Figs. 99 and 101, and the shapes of teeth later.

But the most important consideration in devising these movements is that they shall have no positions for skips, derangements, etc., and that the continued action is absolutely positive throughout. Also that in high speeds or in heavy movements of this kind the driven piece should be given a retarding motion in approaching the stop, and an accelerating movement when it moves off from the stop point, to avoid shocks and breakages.

Here, as in the intermittent motions of Figs. 140 to 143, there may be employed engaging and disengaging spurs and also segments.

In Fig. 145 we have an example where spurs and pins are used in which A is the driver revolving either way continuously, and B

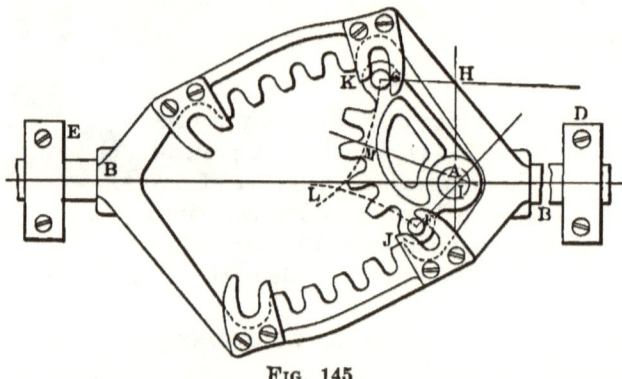

FIG. 145.

the follower sliding in guides D and E forward and back in repetition, a complete movement both ways being made in each revolution of the driver A.

The wheel A and rack frame B are of the same thickness, with teeth to engage each other for the principal movement, the correct construction of which has already been explained.

In the wheel A, at F and G, are pins to be engaged between the spurs FJ and GK made fast to the side or sides of the frame and used to insure the transfer of engagement of teeth from the one side to the other side of the rack frame BB.

The spurs are here made to conform with the cycloidal curves GL and FL in order that the motion of B, when the pins move outward in their slots, may be such as to allow the teeth to go into engagement without interference. The spurs F and G may be extended as far as desired short of interfering with shaft A, but the spurs J and K must be cut away so as to allow the pins to enter the slots, just as they necessarily would if there were nothing but pins and spurs. In this example J and K are cut quite abruptly, K most so; by reason of which quite a shock will occur when the pin G enters its slot and strikes the spur, the intensity of which may be judged of by taking account of the length of the perpendicular AH from the axis A to the normal HG to the spur. The normal FI at F makes the perpendicular distance AI relatively very short, from which it appears that the shock due to contact of the pin at F would be very slight.

To reduce the shock at G, the spurs may be extended to near M, where the normal to the cycloidal curve GML passes near to A. This, however, seriously shortens the segment FG of the segmental wheel A, greatly reducing its number of teeth. The movement would thus be much better adapted to high speeds, and judgment is to be used in designing to obtain the best adaptation to circumstances.

Assuming the velocity of A constant, the velocity of movement

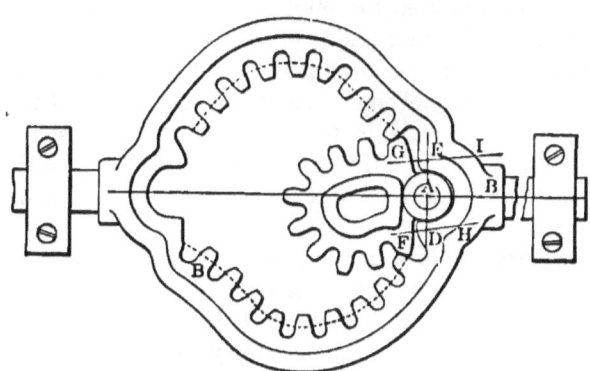

FIG. 146.

of B will be variable, and always proportional to the distance on a perpendicular to BB from A to the intersection with the normal to the spur at G while the latter bears on the spur and to the pitch line while the teeth are in engagement.

This eccentric form of A for alternate motions is objectionable on the score of ease of movement, and may be avoided except where a varied motion is essential. In Fig. 146 the variation of motion is less objectionable than here for a durable high-speed movement.

In Fig. 146 we have an example where solid engaging and disengaging segments instead of pins and spurs are adopted with which to reverse the movement with certainty and positively. As in Figs. 142 and 143, no pieces are attached by screws to become loosened, occasioning derangement.

At reversal of motion the first driving contact occurs at F or G, according to direction of motion. Normals at the contact point being FH or GI, the shock due to contact will be proportional to the length AD or AE, as the case may be, DAE being perpendicular to the direction of motion of BB and corresponding to the line of centers.

The contact points F and G are put as near to the center A as possible to minimize the shock due to initial contact, immediately following which the contact moves outward to the pitch line of the teeth when the latter engage in action and continue the movement.

This movement gives the driven piece B the most rapid motion when near the middle of this stroke, the rate of motion being for each instant always proportional to the length of the perpendicular from A to the contact on the pitch line for that instant.

The ends of the first teeth of both A and B are cut to circle arcs about the center A and of some breadth of bearing, to prevent the disabling of the movement unduly soon by wear of those parts.

The hub of A is made to fit closely into the extreme portions of the opening of B to facilitate bringing A and B into the correct relative positions when the contacts between teeth are shifted from side to side.

Any non-circular form may be given to A, as preferred; though it will be found, on trial of several various curves, that it is favorable for smooth running to have both the curves draw in toward A as in Fig. 146, instead of the contrary, as in the case of one side of A in Fig. 145.

These wheels may be made with conical form of A and of B, the cones having a common vertex more or less remote from A, and the forward and back motion of B occurring as if swinging about an axis through the vertex of the cone and perpendicular to the axis of A.

Also Figs. 145 and 146 may be laid out with BB on a circular curve and the same be put into the spherical form, though a considerable cut and try work, with templets, will probably be required before the final adopted figures are brought out.

The Office of Motion Templets.

In all this work of determining wheels and curves, as in Figs. 140 to 146, templets will be found most useful. They may be formed of paper or cardboard, readily cut out and tried; and again others cut and tried, until the shapes of good working wheels are arrived at, while by any other method, coupled with contempt for templets, there might be total failure.

Second. Unlimited Alternate Motions.

Mangle Wheels and Racks.

The pitch lines of these movements have been considered in Figs. 102 and 103, and the forms of teeth for these pitch lines present no new problems. For the treatment of the various cases of non-circular pitch lines, and teeth for the same which are likely to arise in connection with this subject, we may refer to the principles already given.

For a mangle rack with variable velocity-ratio, the pinion for the same may be non-circular when the rack pitch line is to be curved as in Fig. 39 or Fig. 46, in order that the axis of the pinion A may be more nearly stationary during a movement forward or back.

For a mangle wheel, the pinion may be non-circular and the pitch line wavy that the axis may follow a smooth or non-serrated groove, as in Fig. 102, or 103.

The pitch line on the mangle wheel, besides being wavy, may be varied in general outline from smaller to greater radius, even making several turns around the axis of the driven wheel B and return to the starting-point. In this case the velocity-ratio will vary for the non-circular form of pinion, A, as well as for the wandering form of pitch line on the wheel B. The velocity-ratio in this case will consist of a very complicated cycle of changes.

CHAPTER XII.

TEETH OF CIRCULAR GEARING.

DIRECTIONAL RELATION CONSTANT.   VELOCITY-RATIO CONSTANT.

FIRST: AXES PARALLEL.

No principles in the theory of gearing, beyond those already treated under non-circular wheels, remain to be brought out here, the leading topic for our present study under circular gears being the special features pertaining to circular gearing, and some practical suggestions concerning them.

Circular gearing, though all coming under one general theory in common with non-circular, viz., the development of tooth curves by the rolling or describing templet, may yet be classified and named in two, and possibly three, sections, viz.:

        I. Epicycloidal Gearing.
       II. Involute Gearing.
     III. Conjugate Gearing.

I. EPICYCLOIDAL GEARING.

Here, not only the pitch lines are circles, but the describing circles also, so that the tooth outlines are *epicycloidal curves;* described by a tracing point in the circumference of a rolling circle, or generatrix, as the latter rolls on the periphery of the pitch circle as directrix.

When the rolling is on the outside the generated curve is an epicycloid, and when inside it is an hypocycloid; tooth curves here being therefore always epicycloids and hypocycloids.

SOME PECULIAR PROPERTIES OF EPICYCLOIDS AND HYPOCYCLOIDS.

In Fig. 147 A is a pitch circle upon which the describing circle D rolls carrying the tracing point P and tracing the curve $FPGH$. This curve, thus generated by the rolling of one circle upon another, is called the epicycloid. If the rolling were inside the directing circle A, as in Fig. 148, it is called the hypocycloid.

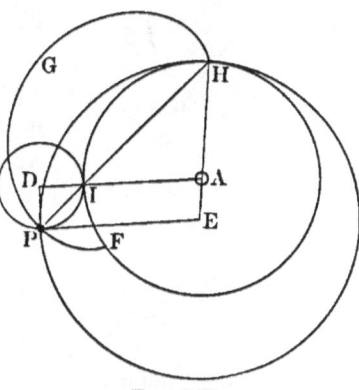

Fig. 147.

One peculiar property of the epicycloid is that it may be generated by the rolling of two different circles upon A; viz., first, PD rolling at I with the tracing point P tracing the epicycloid $FPGH$; and, second, PE rolling at H with a tracing point at P, provided that the diameter of PE, less the diameter of PD, equals the diameter of A. With this relation of diameters, $PD = AE$, $DA = PE$, and triangles PDI, PHE, and IHA are all similar triangles. These conditions remaining permanent during the rolling, P must remain on the line through I and H as instantaneous centers of motion, so that PI or PH remain a normal to the epicycloid at P from which it appears that one and the same epicycloid $FPGH$ will be described by either circle, rolling as described, the difference of the diameters of which circles equals that of the directing circle A.

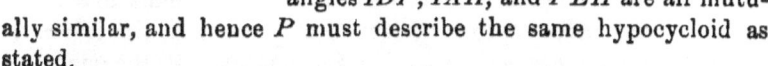

Fig. 148.

Similarly, if two circles as D and E roll inside of the pitch line A, as in Fig. 148, the sum of whose diameters equals that of the pitch line A and carry a tracing point P, they will generate one and the same hypocycloid FPG; for constantly $PD = AE$ and $DA = PE$, and the triangles IDP, IAH, and PEH are all mutually similar, and hence P must describe the same hypocycloid as stated.

When the two circles D and E are equal, as in Fig. 149, the hypocycloid becomes a straight line and is a diameter to A, as in the *White's parallel motion*.

130   PRINCIPLES OF MECHANISM.

Epicycloidal gearing is made in considerable variety, as best adapted to various purposes, and may be classified as follows:

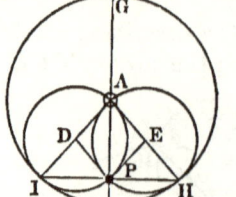

Fig. 149.

1. Flanks radial.
2. Flanks concave.
3. Flanks convex.
4. Interchangeable sets.
5. "Pin gearing."
6. Rack and pinion.
7. Annular wheels.

Treating these separably, we have

1st. Flanks Radial.

In Fig. 150, we have a pair of pitch circles, for wheels to be fitted with epicycloidal teeth having radial flanks, A being the driver, B the follower, and C the point of tangency.

For convenience, we will take C as the origin of all the tooth curves. Then, as in Fig. 112, the describing curve, in this case the circle D, is placed inside of the pitch line D with the tracing point a at C, and rolled, without slipping, to the left along the inside of B, the tracing point a describing the tooth curve Ca. Also the circle D' of the same diameter as D is placed on the outside of A with its tracing point b at C, and rolled toward the

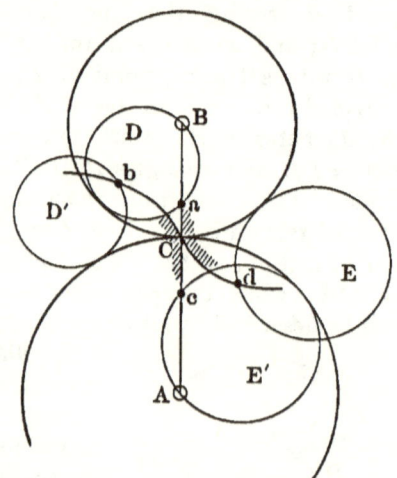

Fig. 150.

left, the tracing point b describing the tooth curve Cb. The circles D and D' must be of the same diameter, or may be the same circle and tracing point, used in succession inside of B and outside of A. The tooth curve Ca is to serve as a flank of a tooth for B and the curve Cb for the face of a tooth of A, the two matting and working correctly together for contacts at the right of ACB. (See Fig. 113 and explanations.)

In like manner the circles, or the same circle, E and E' are rolled once inside of A and once outside of B, with the tracer c or

d starting at C and tracing the mating tooth curves Cc and Cd, the former for a flank of a tooth of A and the latter for a face of a tooth of B, the two furnishing contacts at the left of C, all as explained in Figs. 114 and 115.

We have, now, cCb a contiguous face and flank or a full tooth curve for one side of a tooth of the wheel A, and likewise aCd the one side of a tooth of the wheel B, which teeth mate perfectly, and will work together smoothly transmitting motion from the driver A to the follower B with a constant velocity-ratio of the same value as if the directing or pitch circles A and B were rolling one upon the other without slipping.

The pitch lines being here circles, other tooth curves described with the same describing circles will generate tooth curves which will be copies of those already obtained, as cCb and aCd. Hence it is only necessary to copy these curves as often as needed at the proper pitch spaces around the wheels, when, on describing the addendum and dedendum circles, the drawings of the wheel teeth will be completed, fillets being struck in as hereafter explained.

The leading peculiarity of these teeth is that they have radial flanks due to the fact explained in Fig. 149 that the describing circles have diameters equal half those of the pitch circles within which they roll, and hence the styling *radial flanks* for these wheels.

Following the contact between these teeth from beginning to end, for A turning right-handed driving B, we find that the initial contact between a pair of teeth occurs inside of the pitch circle A to the left of C, and moves on through the point C and ends inside of the pitch circle B at the right of C, and following along an "S" shaped curve; for full discussion of which, see *path of contact*, limit of contact, etc., Fig. 177.

For convenience in the drawing of radial flanked teeth, we see from the above that it is only necessary to draw flanks as radial lines, and faces by using a rolling circle on the outsides of a pitch line, the diameter of which equals the radius of the mating pitch circle.

For proportions of teeth, see Fig. 110 and rules accompanying.

2d. Flanks Concave.

In Fig. 151 we have a pair of pitch lines, *A* and *B*, to be equipped with epicycloidal teeth which have concave flanks, the object of which may be to give to the teeth greater strength than if the flanks were radial. The increased thickening of teeth in the flanks may be judged by noting the departure of the hypocycloid from the radius in Fig. 151.

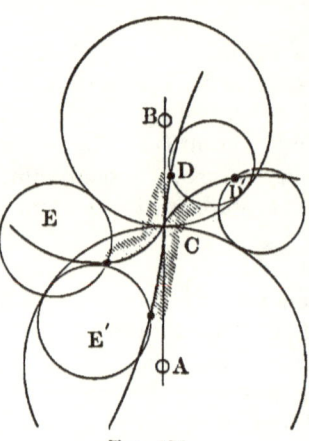

Fig. 151.

The size of the describing circle is arbitrary, and may be chosen to meet the most fastidious taste or fancy as to thickness or strength of tooth, the only restriction being that the diameters of the circles *D* and *D'* be the same, and likewise for *E* and *E'*.

The complete tooth curve for *A*, and also for *B*, is shaded to distinguish it. A templet may be fitted to this curve and mounted to swing around the center, and at every point for a tooth curve it may be held and the tooth curve marked or copied from the templet, as explained among *practical operations*. (See Page 153.)

With the tooth curves marked off around the wheels, the addendum and dedendum circles drawn in, and the root fillets struck, the tooth outlines for the wheels will be completed.

3d. Flanks Convex.

In Fig. 152 the rolling circles have diameters which are greater than the radii of the wheels within which they roll, as in the case of the wheel *E* in Fig. 128, or *J*, Fig. 119, giving flanks of teeth that are convex. The finished teeth for this case will be found very weak, as evident from the undercut appearance which the teeth present. The filleting at the root will help this, though the teeth are still weaker than in the previous cases, but will be found of favorable form in light-running machinery.

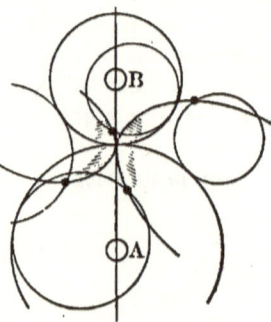

Fig. 152.

4th. Interchangeable Sets of Gears.

These wheels are called interchangeable because, of any number of them of various sizes having the same pitch and describing circle, any two will work together correctly. This is seen not to be the case with Figs. 150 to 152.

The essential requirement for this interchangeability is simply that the generating circles D, D', E, and E'' be always of the same diameter for generating faces and flanks of teeth throughout, as shown in the circles D, D, etc., in Fig. 153.

For instance, A and B is a pair of wheels differing in size and selected at random from the set, neither being the largest nor smallest, for which we have a pair of tooth curves in contact at C, shaded as shown, the one aCd belonging to wheel A, another bCc to wheel B, etc.

Now, the curve aCd is throughout described by the rolling circle D, and, as the latter is chosen independently of BB', etc., it appears that this tooth curve aCd is entirely free of all reference to any other wheel of the set, and only dependent upon the pitch line A and describing circle D. The same is true of any other wheel of the set, as A', B, B', etc., the tooth curve of any one wheel being peculiar to that wheel only, since the describing circle is one and the same throughout the set.

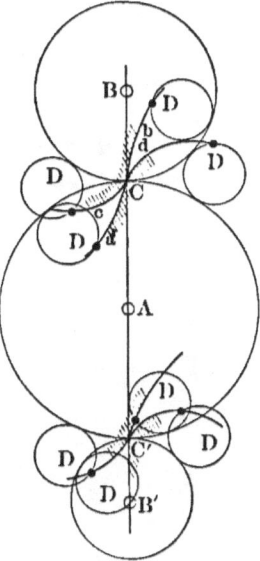

FIG. 153.

Hence, to realize an interchangeable set of epicycloidal gear wheels, where any two whatever will mate correctly, it is only necessary, in addition to a constant pitch, that the describing circle, generating the teeth, be one and the same throughout. The "change wheels" of engine lathes in machine shops is a well-known example of gears of this kind; that is, in "sets."

In Fig. 153, three wheels of a set of this kind are shown with tooth curves drawn where the describing circle is in common for all.

At the size B' the flanks are radial, as in Fig. 150, because here we strike the size where the radius of B' equals the diameter of the describing circle D. For wheels of this set smaller than B' the flanks would be convex, as in Fig. 152.

134   PRINCIPLES OF MECHANISM.

As the wheels of the set are made larger, that one with infinite radius becomes a rack, where the curves for faces are identical with those for flanks.

If we carry the variation of curvature still further in the same direction, the rim of the wheel becomes concave, and the smaller wheel of the pair finds its place inside the larger. This is sometimes called internal or annular gearing. In this case a given annular wheel has tooth curves which are exactly the same as those for a wheel of the same diameter with its mating gear in outside action.

In the well-known example of the "change wheels" of engine lathes, the gears and teeth are small, and correct teeth are not so

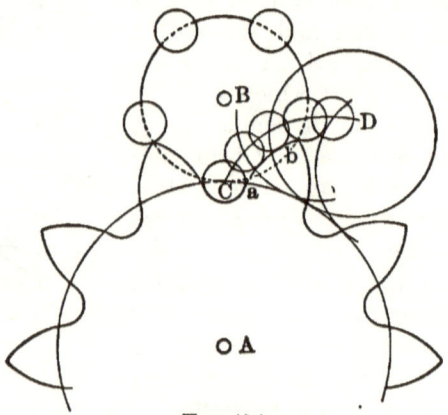

FIG. 154.

important as in sets of patterns for cast-iron mill gearing with larger teeth, where the manufacturer must be prepared to furnish pairs of gears of almost any sizes with correct teeth. The best way to meet this demand is to have patterns made in sets, as above explained, any two of which will mate correctly; then with all of this set an occasional new pattern will work correctly, one new pattern thus providing for many new pairs.

5th. **Pin Gearing.**

This is called pin gearing because the teeth of one wheel are pins and of the other are spurs to engage the pins, as shown in Fig. 154.

To draw correct working gearing of this kind, take A and B as the pitch lines, and the circle at C as the section of a cylindrical pin or tooth. Draw an epicycloid CD, with B as the describing circle.

CIRCULAR GEARING.

This epicycloid will be the path of the center of the pin C, on the supposition that the pin is made fast to the circle B, and that the latter is rolled, without slipping, along the pitch line A, as shown.

The circles along the epicycloid CD represent the pin tooth in various positions agreeably with corresponding positions of the pitch line B, in its rolling, and the curve ab, just tangent to these circles of the pin, will be found a correct tooth curve for one side of a tooth or spur of the wheel A, by aid of which all the teeth of the wheel A may be drawn. The pair of gear wheels A and B may therefore be fully drawn, the pins and teeth being properly distributed in pitch.

A brief examination of Fig. 155 will show that the contacts between these teeth and spurs are on one side of the line of centers AB, receding if A is driver, and approaching if B is driver. Also, it is found that the receding contacts are the most efficient, as clearly evinced by the extreme case of Fig. 156, where, with A for driver, the extended tooth curve acts like a wedge to force the pin along in its path, while if B is driver the pin is forced down upon the tooth curve of A, the latter serving very much to block or to hinder the movement of B; and when, as in practice, friction between the pin and tooth is considered, the problem is still more prejudiced.

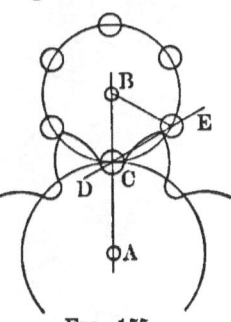

FIG. 155.

Fig. 156 is an extreme case to illustrate more forcibly, but in a more moderate one, as in Fig. 155, the principle still holds to a certain extent.

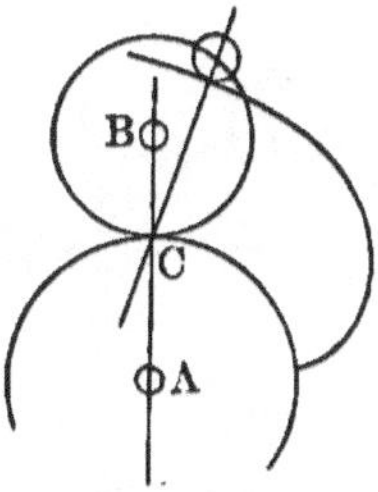

FIG. 156.

As tooth contacts are on one side of the line of centers, the arc of contact is considerably limited, and each problem should be examined with care to determine whether one pair of teeth quit contact before the next succeeding pair commence it, the latter usually occurring when the center of the pin is some considerable distance past the line of centers. (See *Mac-Cord's Kinematics*, page 210.) For large pins, the amount to allow for this is greatest, and vanishes as the pin diameter becomes zero.

In some examples these pin teeth have been made as rollers on smaller pins, with the view of reducing the friction.

Pin gearing has been quite extensively used in the cheaper brass clocks which have flooded the country within the last forty years. In these the wheels are drivers, as seen above to be advisable, and made of sheet brass, while the pinions have pins of straight and smooth steel wire, held at both ends in holes in a pair of heads or disks of brass.

Inside Pin Gearing.

Pin gearing is applicable in inside gearing, the curve CD, Fig. 154, instead of being outside would be inside, and consequently an hypocycloid, but the same principles apply as before.

In the special case for inside gearing, where the pinion is half the size of the wheel, the pins of the pinions move along certain diameters of the wheel, according to Fig. 149, and sliding blocks in grooves may be introduced, with holes to receive the pins of the pinion as in Fig. 157.

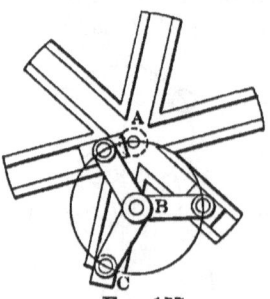

Fig. 157.

One of the arms of B may be cut away and a grooved arm of A also, and still the movement would work complete. The remaining two arms of B could be placed at any other angle with each other as 45°, 87°, etc., provided the corresponding two grooved arms of A were at half the angle intervening, and the action would be perfect.

Pinion of Two Teeth.

Approximately, rectangular teeth may be employed instead of cylindric, as shown in Fig. 158, where the longer sides of the pins are radial with B when, as shown by Fig. 149, the circle BEF as a templet rolling on the circle AF, a tracing point at E would describe the epicycloid KDE, while if rolled inside of BF it would describe a radial side of the tooth or pin E for the curve KDE to work against.

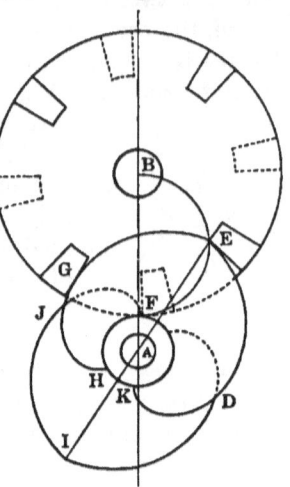

Fig. 158.

A is here the pinion and has only two teeth, the least number possible, though it may have as many more as desired.

The construction is a little peculiar, the teeth of *B* alternating upon opposite sides of a central disk, and for *A* we have two heart-shaped pieces, *KDEGH* and *FJID*, at such distance apart on a hub that the disk of *B* may work between. Then the contacts will alternate from side to side as the rotation proceeds. The part *AEG* will just fill the space between the two teeth *E* and *G*, and the wheels may work either way.

It is hardly practical to make *B* driver here, since the common normal at *E* passes so near to *A* that, considering friction, *A* would serve as a block to hinder the driving action of *B*.

In Fig. 159 we have a two-leaved pinion *A* driving *B* as equipped with cylindric pins, the construction being the same as in Fig. 154, except carried to the extreme case of two teeth for *A*.

Care should be exercised that a beginning contact at *D* is fairly made by the time a preceding contact ceases at *E*. The rolling of the pitch line *B* as describing templet, on the pitch line *AC*, gives rise to the epicycloid dotted in at *FG*, parallel to which and at a pin radius *EG* from it is drawn the tooth curve for the pinion *A* as shown, as in Fig. 154.

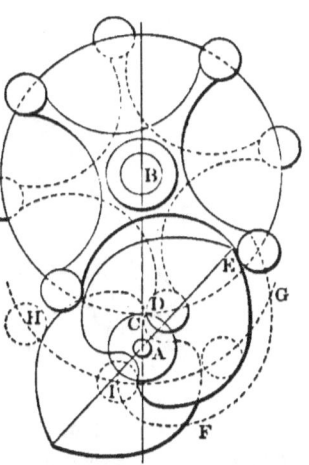

Fig. 159.

For inside gearing, the same pinion *A* may be placed *inside* the same wheel *B*, when the pitch line of *B* will fall at *HI*, tangent to *A*; and the pins will have contacts at *H* and *I* as shown, the center of *B* being at *B'*, and below *B* by the amount 2*AC*.

Rack and Pinion.

For the case of the rack and pinion, the latter may have either the pins or the teeth, the same principles applying.

Wheels of Least Crowding and Friction.

A peculiar form of pin gearing is obtained by use of a describing circle of somewhat unusual size, as in Fig. 160, the pin having its side, which is within the pitch circle, a complete hypocycloid.

Here the common normal makes the least possible maximum obliquity with the common tangent to the pitch lines, when the

Fig. 160.

teeth are made just so large that the arc of contact embraces but one tooth contact at once; this obliquity being but about half that for circular pins when the contacts must all be on one and the same side of the line of centers, as in Fig. 154, and also only about half the obliquity of teeth in Fig. 150.

These wheels of all others will have the least possible crowding action or tendency to force the axes from each other when made as just specified, so that the arc of contact but slightly exceeds the pitch, the friction due to crowding being therefore at a minimum, and the pressure to turn the driven wheel will be at a minimum for two reasons: first, because the driving pressure between teeth is nearly tangent to the driven wheel, and, second, because this last-named pressure is to be compounded with the least possible crowding pressure.

But the teeth of these wheels are unusually weak, and it can be recommended for only light-running machinery.

On account of the slight obliquity of the line of action with the common tangent to the pitch lines, a variation of the distance between the centers A and B would entail less prejudicial action than in Figs. 150 to 159, so that in Fig. 160 we find a suggestion for such wheels as used on clothes-wringers, straw-cutters, etc., when the distance between A and B is varied in a very large ratio.

6th. The Rack and Pinion.

In any of the preceding cases for Figs. 150 to 160, we obtain a rack and pinion by giving to one of the wheels of the pair an infinite radius.

For the case of radial flanks, as in Fig. 150, if we make A infinite, we obtain Fig. 161.

Here the radial flank of A will be simply a perpendicular to the straight line AC, while the face curve CD for B will in this particular case be an involute to B, since the describing circle to roll inside of A, which is infinite, will be in-

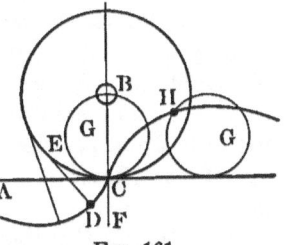

Fig. 161.

finite also; ED being one position of that circle or straight line, carrying the tracer D, and tracing the involute, CD, as the rolling advances, along CE. Thus the line DE will equal the arc CE, and the involute tooth face CD can readily be drawn.

For the radial flank of B and face of A the describing circle GH must be half as large as B, which, with a tracer H, rolled on AC will in this particular case trace a cycloid, CH.

Then we will have FCH for a full tooth curve for A, and BCD a full tooth curve for B; and the correct teeth can be constructed.

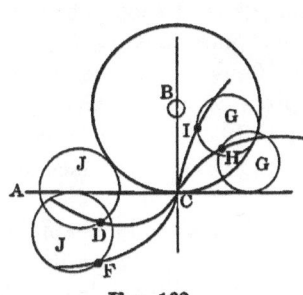

Fig. 162.

As before, CD and CF will be mating tooth curves, having a working contact to the right of C, while the mating tooth curves CB and CH will have contact at the left of C.

One peculiarity of the action between CD and CF is that only the point C of the line CF will go into action with the whole of CD, and, as a consequence, the rack teeth would, in practice, become unduly worn at C.

For the case of concave flanks, the describing circle G, Fig. 162, must have a diameter less than the radius of B to describe the mating tooth curves CI and CH; and any circle, J, less than infinity will serve to describe the mating tooth curves CD and CF. The line FCH is a tooth profile for the rack AC, and the line DCI is a tooth profile for the pinion B. With these the teeth can be fully drawn for the rack and pinion.

The curves CF and CH will be cycloids, with bases on the line AC.

For the case of flanks convex, the describing circle, as in Fig. 152, to roll inside of B, must have a diameter greater than the radius of B; but for the rack, convex cycloidal flanks are impossible for the reason that the straight-line flank of Fig. 161 is struck with an infinite circle, the largest possible. Also, it is seen that by rolling any describing circle along CA, from C towards A, the line CF could not be described as downward toward the right, for the reason that no instantaneous center on CA will serve for striking a curve in such position. The straight line CF, normal to CA, Fig. 161, is therefore the extreme case of carrying F toward the right.

The Case of the Rack in Interchangeable Sets has already been treated in connection with Fig. 153, also *Pin Gearing*, in connection with Figs. 155 and 159.

7th. Annular Wheels.

In Fig. 163 we have the general case, the smaller pitch line being within that of the annular wheel. The circle carrying the tracers H or I is rolled inside of A, and also inside of B, to generate a face of a tooth of A and a flank of B. These curves are both hypocycloids; but, according to the general theory, are seen to be the proper ones to work together for a face and flank. For the outside curves the circle J carries the tracer D or F to describe the epicycloids CF and CD as mating curves for a face of B and flank of A.

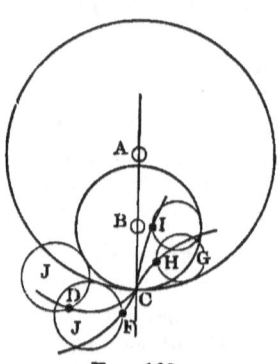

FIG. 163.

Both sets of flanks are here concave, and that for A must always be so; though that for B may be radial or convex by using a larger describing circle to carry the tracers H and I.

CIRCULAR GEARING.

A peculiar case of interference of the teeth of the annular wheel and its pinion occurs when the sum of the diameters of the describing circles G and J exceeds the difference of diameters of A and B, as discovered by A. K. Mansfield and Prof. C. W. MacCord.

The limit occurs when the added diameters of the describing circles G and J just equals the difference of diameters of the pitch lines A and B, as shown in Fig. 164. At this limit the faces of the teeth of the internal gear A have correct acting tooth contacts with the faces of the teeth of the inside pinion B, as may be proved by aid of the alternate describing circle K.

Referring to Fig. 148, it is seen that the hypocycloid DL, Fig. 164, may be regarded as being now traced by either G or K carrying the tracer D; and that aDC is a straight line. Also referring to Fig. 147, it is clear that the epicycloid DM may be regarded as in the act of being traced by either J or K carrying the tracer

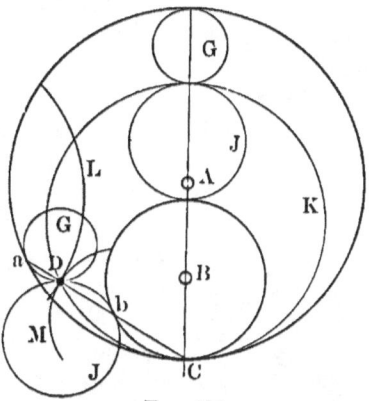

Fig. 164.

D, and that the points DbC are in a straight line. Hence, $aDbC$ are points in one and the same straight line; and the hypocycloid DL and epicycloid DM are tangent to each other at D, as required for tooth contacts; and we have, at this limit, in addition to the usual face and flank contacts, this face to face contact at D.

If $D + J$ is larger there will be interference at D between faces, while if smaller this face to face contact is lost.

In this particular case we have the advantage over ordinary gears of an additional and unusual contact at D, and a most excellent one, except for its remoteness from C, where the amount of slip of face surfaces over each other is greater than in the other and usual contacts.

From what has been said, it is plain that the describing circles G and J can be varied, either smaller than the other, but that they must not be larger than that they will fit in between the two pitch circles, as shown, under penalty of positive interference of the teeth at D, but that they may be smaller than shown, with loss of contacts at D.

Wheels in sets may be worked in annular gearing, that is, any of a number of pinions may work correctly in any of a number of annular wheels, provided that one and the same describing circle be used throughout, and that the facts of interference be observed, as pointed out above.

Pin annular gearing may also be used by following the principles explained under pin gearing, and the pins may be in either the wheel or pinion.

II. INVOLUTE GEARING.

As explained in connection with Figs. 119 and 121, logarithmic spirals may be employed for describing curves which, rolled on the pitch lines A and B, inside and outside in the usual way, will develop faces and flanks by pairs, the curves thus traced being involutes to circles somewhat smaller than the pitch lines.

Thus, in Fig. 165, take the pitch line $A—DCH$ and roll the

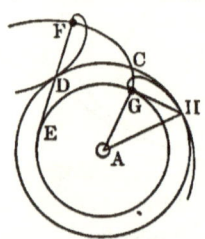

FIG. 165.

logarithmic spiral FD from C as shown, with a tracer at the pole F. The curve CF will be an involute to a circle EG, inside of A, to which the line EDF is a tangent. This is made evident by the fact that the logarithmic spiral is a curve of constant obliquity, that is, that any line or radius vector, FD, makes a constant angle with a tangent to the spiral at D, and, consequently, a constant angle with the tangent to the circle at D, and also with a radius to the circle at D. Hence, FDE will always be tangent to some one circle EGA; and it is plain that we will obtain one and the same curve, CF, whether the latter be generated by the rolling of the spiral, with tracer F, along CD, or by the unwinding of the cord FE, with tracer F, from the circle GE. This curve, CF, will thus be an involute to the circle AGE. This curve as an involute starts at G, while as developed by FD it seems to start at C. To supply CG by the rolling spiral, place the pole and tracer at C, and roll inside along CH as shown, when CG will be traced just to G and stop; for then the radius of curvature of the spiral at H will be HA, since to construct this radius of curvature for a logarithmic spiral draw the right-angled triangle AGH, when AH is that radius. The spiral then cannot roll beyond H, and the curve CG stops at G.

As tooth curves, CF is to serve as a face and CG as a flank, and, as developed by the spiral, are seen to fall within the general theory of development by rolling curves. But, in practice, for convenience, the curves are drawn as involutes to the circle GE, called the base circle or base of the involutes.

When the teeth need to be cut deeper between than CG, such portion is cut away arbitrarily below EG as shall make room for the interengaging teeth.

To Obtain Tooth Curves for Wheels with Involute Teeth.—Draw the pitch circles A and B in common tangency at C, Fig. 166, then a straight line DCE through the point of common tangency or pitch point C, and perpendiculars AD and BE from the centers A and B upon this line. Then, with AD and BE as radii, draw the base circles FD and HE of the involutes. The involute may be drawn by unwinding a thread from the base circle, the thread having the tracer attached, or by drawing several tangent lines as shown, and stepping with dividers from C to D, then with the same number of steps to F, then back to other points of tangency and out on the tangent to points G, etc., for as many points as desired, when the involute FCG may be drawn in; likewise for HCI.

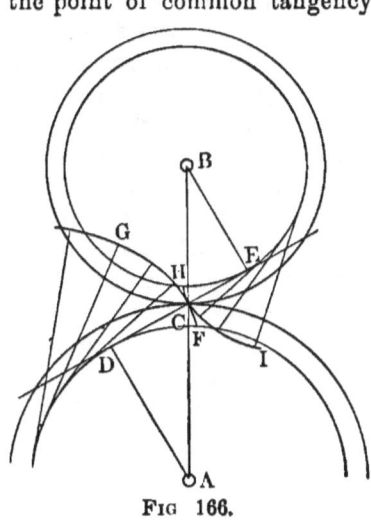

Fig 166.

This gives us a pair of tooth curves from which the teeth may be drawn in, clearance room within the base circles DF and EH being assumed. As the finished gear wheels revolve, the tooth contacts all occur on the line DE, remaining on and moving along it; and there can be no correct contacts outside of the length DE, though between D and E there may be several. Attempted contacts beyond D or E give rise to serious interference.

In practice it is rare that the contacts extend to the whole of DE, though they may, by laying out the wheel with that result in view, as by making the addendum circles cut at D and E.

These gear wheels possess the peculiar property that the distance between centers A and B may be varied at pleasure, without altering the correctness of action of the teeth upon each other.

For, examining Fig. 166, we find that the triangles ACD and BCE are similar, so that the pitch point C is always at the same proportional position between A and B, whatever the magnitude of AB, allowing AD and BE to remain constant. Under these conditions any number of correct working tooth outlines may be drawn with varied distances AB. The spaces between the teeth along on the line FD, or on DE, or on HE all equal the *normal pitch*, which is constant; while the spaces between the teeth along on the pitch circles equal the circumferential pitch. This latter differs slightly with the normal pitch, and is slightly variable if we regard the pitch circles as remaining tangent to each other at C while the line of centers AB varies.

This gearing is often recommended for situations where the distance AB, through carelessness, wear of parts or otherwise is subject to changes; and where epicycloidal gearing, requiring AB to remain constant, will not serve correctly.

These involute wheels in a group of various sizes of the same normal pitch will all work together interchangeably, and they are probably the best for change gears for engine lathes and the like.

In the Rack and Pinion the wheel A may be regarded as of infinite diameter when the pitch line becomes the straight line AC, Fig. 167. The base circle also becomes infinite and at an infinite distance from C, so that the line CD is infinite and the tooth curve FG for the rack becomes a straight line perpendicular to DE, the face and flank being both parts of one and the same straight line. The tooth curve, HI, remains the same as before, for a like inclination of DE. The addendum line for the rack should never go above E.

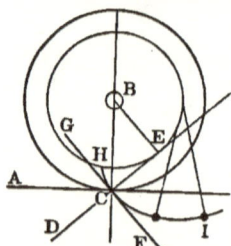

Fig. 167.

For the Annular Wheel and Pinion, the tooth curves for the former become concave as at FG, Fig. 168, while the pinion remains the same as before for the same inclination of CED; and for this a group of pinions of the same pitch will interchange with this wheel.

In the present case there can be no contacts between D and E, but may extend from E through C to infinity, except for interference at L.

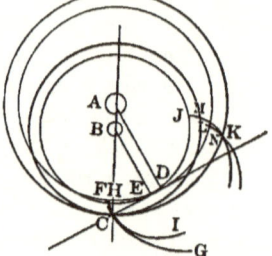

Fig. 168.

In these gears, as well as in epicycloidal, there may be interference of teeth, as shown in an exaggerated view at KL and KM, the tooth curves intersecting at K. This is most likely to occur where the pinion is relatively large. To examine a case for interference, draw the pitch circles and on them step off equal pitch distances from C to M and N, and through these points draw the involute tooth curves, as shown. If these curves do not intersect, as at K, within the addendum of the pinion and the circle DL of the wheel, there will be no interference at that position, and, if the test is applied at the intersection of the addendum circle for B with the base circle to AD, with no intersection of tooth curves, there will be no interference of teeth for the wheels. Modifications of amount of interference are affected by varying the addenda and the inclination of CED.

Different systems and sizes of teeth may be combined in a single gear, as illustrated in Fig. 169 of a pair of cir-

FIG. 169.

cular gears, 2 to 1, velocity-ratio constant, embracing both the epicycloidal and the involute forms of teeth.

III. CONJUGATE GEARING.

This term is applied to teeth where the tooth of one wheel is assumed and the mating tooth for the other wheel found from it, usually by the practical operation of templets, as foreshadowed in Fig. 108 and more definitely shown in Fig. 122, which may be followed fully here with circular pitch lines.

SOME PARTICULAR CASES.

1st. Flanks Parallel Straight Lines.

In Fig. 170, A and B are the pitch circles somewhat removed from tangency to better show the work. One of the parallel flanks is CG. From C step off equal spaces, marked 1, 2, 3, 4, etc., on

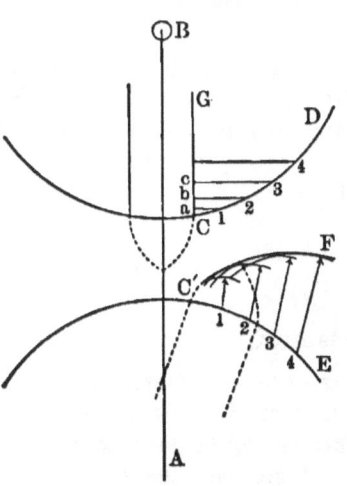

FIG. 170.

the pitch line of B. Likewise from C' step off the same equal spaces on the pitch line A as shown, C and C' being a half-tooth thickness from the line of centers. Draw $1a$, $2b$, $3c$, etc., perpendicular to CG. Then with $1a$, from wheel B as a radius, draw a circle arc about the point 1 on the wheel A; also, with $b2$ as a radius, draw a circle at 2 on A. Likewise with $b3$, etc., for as many points as required. Then draw an enveloping curve FC', which will be a correct face curve for A, to work on the flank CG as mating tooth curves. Other pairs of mating face and flank lines may be drawn, and the teeth completed, as shown in dotted lines.

Examining the mating curves drawn, it is plain, for example, that the radius from 3 to where its circle is tangent to the envelope $C'F$ is a normal to $C'F$, so that when the points 3 of pitch line A and B are tangent to each other this normal will coincide with the normal $3c$, and c will be the point of contact and tangency between the tooth lines. Likewise for other points, so that the face and flank will slide smoothly with each other throughout, while the pitch lines mutually roll, thus answering to required conditions for tooth action.

2d. Flanks Convergent Straight Lines.

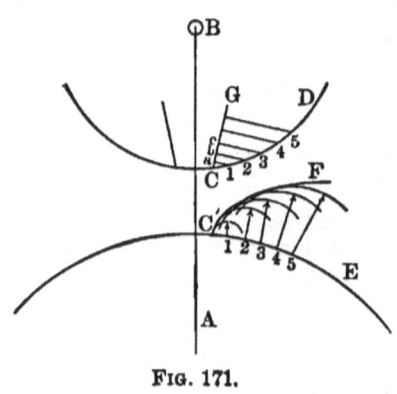

Fig. 171.

In Fig. 171 draw the pitch lines and the convergent flanks CG. Step the equal spaces 1, 2, 3, 4, etc., draw the lines $a1$, $b2$, etc., normal to the flank line CG, use these latter as radii, and draw circle arcs as before on the pitch line A. The envelope $C''F$ will be the proper tooth face for the wheel A, to work on on the flank CG, and the complete tooth can be drawn as in Fig. 170.

3d. Flanks Circle Arcs.

Draw the pitch circles A and B, the line of centers, and a tangent HI at H, Fig. 172. From a point I on HI draw the circle arc CG for the flank of a tooth for B. Through equidistant points 1, 2, 3, etc., draw radii from I. Then $a1$, $b2$, etc., are the radii to use at points 1, 2, 3, etc., on A as centers of arcs, tangent to which

CIRCULAR GEARING.

draw the envelope $C'F$ for a face of a tooth of A that will work correctly on the flank CG.

As another example, the center I may be taken anywhere below the line III, but no higher above than to be on a tangent to the pitch circle B at the point C, in which case CG will be normal to the pitch line. No example is to be found of a correct flank line which is undercut as by placing I still higher.

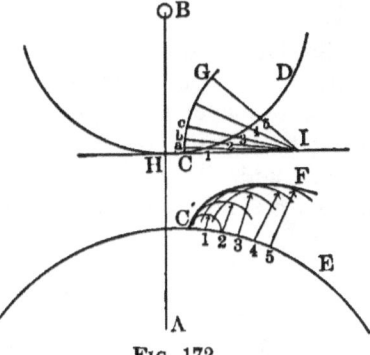

Fig. 172.

Again, in another example, the line CG may be a parabola, ellipse, or other curve to suit the fancy, always being so drawn as not to undercut at C, and in every case the lines $a1$, $b2$, etc., being drawn normal to the curve CG.

The inverse of these cases is practicable where the face of a tooth of A is assumed to be a circle, parabola, or other curve, and the flank found to match it in a similar way.

CHAPTER XIII.

PRACTICAL CONSIDERATIONS.

So far in circular gearing, we have dwelt upon the *theory* of tooth curves which are perfect in action. But the demands of practice are not always so severe as to require the teeth to be thus accurate. There are other considerations also to be noted, such as proper addenda, clearance, fillets, obliquity of action, overlap of action, practical methods, approximate teeth, etc.

ADDENDA AND CLEARANCE.

At Fig. 110 is given a set of rules for proportioning the teeth, which are among the simplest of quite a variety. In cast gearing more clearance is required than in machine-dressed teeth, and more as cast at some foundries than at others, according to care in moulding.

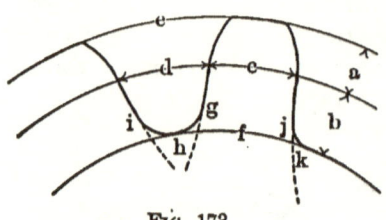

Fig. 173.

The rule for circumferential pitch at Fig. 110 will answer for the average of care in moulding, and for ordinary sizes. For very large teeth the proportion for clearance is too great, while for very small cast teeth it will often be found too small, so that the best rule will probably be the one stated varied to suit the judgment of the draftsman in each particular case.

The rule as to diametral pitch is best adapted for cut or machine dressed gear teeth, where a comparatively slight bottom clearance is sufficient and a much less side clearance. The Brown & Sharpe cutters and instructions are apropos here.

148

In Fig. 173, for cast teeth, cd is the pitch circle, e the addendum circle, f the dedendum or root circle; c being equal to 5/11, d equal to 6/11, a equal to 3/10, and b equal to 4/10 the pitch, according to the rule of Prof. Willis, as given at Fig. 110.

To Strengthen the Teeth, root fillets are generally struck in as at jk, which only extend to a part of the bottom surface at k and up to a comparatively small portion of the flank at j. In most cases the bottom of the space may be limited by a circle arc, ghi, tangent to the flank lines at g and i, and to the root circle, thus simplifying the shape of the bottom of the space and greatly strengthening the teeth. To test this matter the so-called "clearing curve" may be drawn—a curve that would be described by the corner of the point of the tooth, as in Fig. 174.

To Find the Possible Clearing Curve described by the tooth corner b, Fig. 174, as it enters the space def, the side ab acting against the side de, make equal spaces 1, 2, 3, etc., then with $b1$ for a radius describe a circle arc from the point 1 near d; again, with $b2$ for a radius describe the arc from 2 in the space near d, etc., for as many points and arcs as required. Then the enveloping curve to these arcs, as shown, is the possible clearing curve. This curve, of course, extends only the addendum distance below the pitch line. Now a circle arc, drawn in as ghi, Fig. 173, where $a = 3/10$ and $b = 4/10$, will usually, if not always,

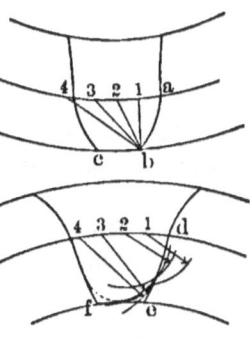

FIG. 174.

be found to go outside of this possible clearing curve, and, as it is the simplest form of outline for the bottom of the space as well as gives greater strength of tooth than does the smaller fillets jk, Fig. 173, it is recommended. Indeed, if ever found necessary, it is usually advisable to go a little deeper to get in the circle arc rather than adopt two smaller arcs. The circle arc is dotted in, and is seen to go considerably below the curve enveloping the circles struck from 2, 3, etc., Fig. 174.

THE PATH OF CONTACT.

The Path of Contact between a pair of operating teeth of **mated** gear wheels is the line followed by the touching point between the pair of teeth, and differs in form in different systems of gearing.

In Epicycloidal Teeth it is the Describing Circle Itself, in a central position, or that of common tangency with both pitch circles, as at C, Fig. 175, where CF is the describing circle. This may be shown by supposing CD equal to CE, and that the describing circle has rolled from where the tracing point F was at D over to the position shown, and likewise from E over to the same position. The rolled curves DF and EF will be tangent to each other at

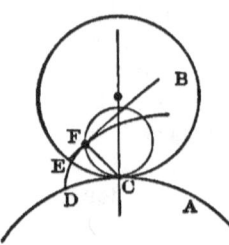

Fig. 175.

F, because CF is a normal to both curves, since C is an instantaneous axis of motion of the tracing point. Therefore F, the point of contact of the tooth curves DF and EF is on the describing circle. Hence the point of contact F follows along on the describing circle CF from C to F, the tooth curves engaging at C and move toward the position shown, making the describing circle CF the path of contact.

In Involute Gearing, it is plainly seen from Fig. 166 that the path of contact is on the line DCE.

In Other Cases, the Path of Contact may Readily be Found. To illustrate, let us find it for the teeth of Fig. 172 as shown in Fig. 176. Draw radii from B to the points 1, 2, 3, where normals to the flank curve Cc cut the pitch circle. Draw circle arcs through the points a, b, c, where the normals meet the flank curve, and extend them to meet the radii from B.

Then, evidently, the flank curve is in contact at a, with its mating face,

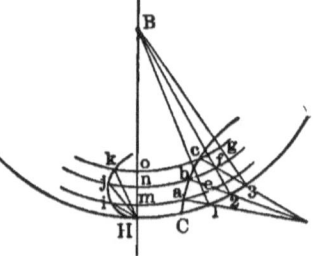

Fig. 176.

when the point 1 is at H on the line of centers, and a at i, where $im = ae$; thus making i one point in the path of contact. Likewise, $jn = bf$, $ko = cg$, etc.; thus giving the curve $Hijk$ as the path of contact for the pair of tooth curves of Fig. 172 as above the pitch lines. Usually the path extends below the point of common tangency of the pitch lines toward A, for the mating tooth curves below the pitch lines as well as above toward B, and this path is always a curve, except in involute teeth.

By the Inverse of this the Path of Contact may be Assumed, and the tooth curves, face and flank, may be found therefrom, as was

PRACTICAL CONSIDERATIONS. 151

done by Prof. Edward Sang, and explained in his remarkable treatment of the subject of gear teeth published some years ago.

The Path of Contact is Limited in Practice where the addendum circles cut the path lines. Thus, in Fig. 177, draw the addendum circles EF and DF. Then the contacts for the epicycloidal teeth of the wheels A and B must begin and end on the S-shaped line or limited path DCE. One pair of teeth should engage at D as soon as the preceding pair quit contact at E, but usually it is sooner. In involute teeth, the contact cannot possibly extend beyond the points D and E, Fig. 166, so that the addendum circles should not overreach those points as has been stated.

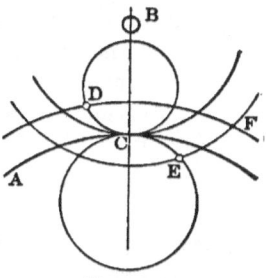

FIG. 177.

The Acting Part of the Flank of a tooth, that is, the entire portion that engages with its mating face, is simply that which extends from the point D to the pitch line of B, or from E to the pitch circle of A, Fig. 177. This portion of the flank is much shorter, usually, than its mating face, thus giving cause for more rapid wear on the flank than face, from which it appears that gear teeth will wear out of shape and not into it, so that if teeth are ever correct in form, they must be made so.

The Line of Action of the Pressures of Contact between the teeth, neglecting friction, is always in the line of the normal, as CF, Fig. 175; this line being therefore properly called the line of action. In involute gearing it is always one and the same straight line, DCE, Fig. 166, but in all other forms of teeth it is a line of varying inclination, like a straight line, iH, jH, kH, etc., Fig. 176.

The tendency of the teeth to crowd the wheels apart depends upon an intermediate obliquity of this line of action, the same being reduced as the maximum obliquity is reduced for a given system of teeth.

Thus in epicycloidal gears, the larger the describing circles, CD and CE, Fig. 177, the less the obliquity and the tendency to crowd. The maximum obliquity of the line of action with the common tangent to the pitch lines is that of the straight line through D and C on the one side, and the line EC on the other.

The Blocking Tendency.

The tendency of a pair of teeth to "block" or stop the wheels is greater in approaching action than receding, and increases with the obliquity of action; so that in wheels of few teeth the maximum obliquity of the line of action should be kept as small as possible, as already explained under non-circular gearing, and sometimes it may be advisable to make the addenda unequal to favor the obliquity in approaching action.

Unsymmetrical Teeth.

These may be made when desired, by using smaller describing circles for one side of a tooth than for the other, for epicycloidal teeth, or greater obliquity of the line of action in involute, etc. Fig. 169 illustrates several such teeth.

To Draw the Tooth Curves in Practice.

Epicycloidal Tooth Curves may be drawn as in Fig. 178, where A and B represent a pair of pitch circles for which it is desired to draw a pair of tooth curves of this form for teeth with concave flanks, the work all being completed on the drawing-board and with only the ordinary drawing instruments. Having the pitch circles drawn at a distance apart, if need be, for perspicuity, adopt some diameter of describing circle $CH = C'H$, such as has a diameter less than the radius $C'B$ as for concave flanks for B. Then, with the dividers, describe the circle in several positions on A, and in B, tangent to the pitch circles as shown. Then with C as the origin of a tooth curve, step off with the spacing dividers on A, as at abc, etc., to near the point of tangency with A of some drawn circle F, say the first one, and then, without lifting the dividers from the paper, step backwards on this circle an equal number of steps to F, and note the exact point. Then beginning at C again, step along on A to near the tangency with the next

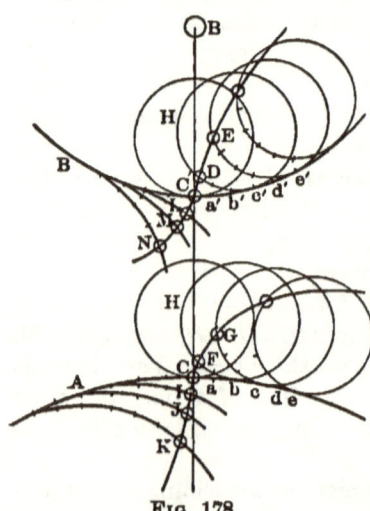

Fig. 178.

circle, and then back on that circle an equal number of steps to G, and carefully note that point. So proceed until the desired number and frequency of noted points is obtained. Then the epicycloid, CFG, may be drawn in through the points noted by aid of an irregular curve.

In stepping from C along on A it is not necessary to come out exactly at the point of tangency with the circle F, G, etc., but the nearest half-step or less from it, and then turn back on that circle to the point F or G, as the case may be. Neither is it necessary, as F is noted, to go back to C, but without lifting the dividers we may retrace the steps from F to A and thence along A to near the next point of tangency and then backwards on that next circle, to G, etc.

Proceed likewise for all the tooth curves to be drawn, as, for instance, the mating flank for C, F, G, using the same describing circle, resulting in the points C'', D, E, etc., in B. Also a face for B in the points L, M, N, etc., and its mating flank I, J, K in A, using the same describing circle for both. We then have a complete tooth profile, $GFCIJ$, for A, and another, $NLCDE$, for B. Templets may be formed to these curves and used to finish drawing all the teeth of the gear wheel A or B, either on the drawing-board or on the

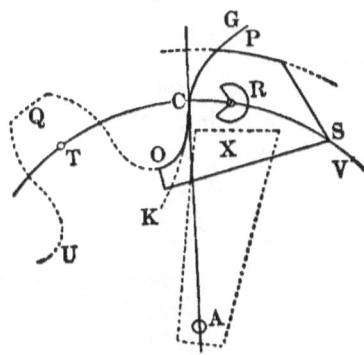

Fig. 179.

wooden pattern being made for use in casting. When the flank is ready for it prepare a templet of cardboard, thin piece of wood, or zinc plate. The latter may be blackened with a chemical as an excellent preparation for visibility of lines drawn upon it. Lay off the curve GCK, for instance, with accuracy. Also, if desired, the addendum circle at P, Fig. 179, and the root circle at O, to fit the root curve of Fig. 173. At R cut an opening with a sharp point R coming exactly to one of the pitch points T, R, V, so that CR

equals a half-tooth thickness. Also at S cut the templet to a point that shall just fit upon the pitch circle. This gives us the required tooth templet. Then, when this templet $OCPRS$ is placed at any pitch point, as R, the edge of the templet PCO is in just the position for drawing a scriber along the edge of the templet to trace the whole tooth profile, including addendum and root curve. Then the templet is to be shifted along the pitch line till the point of the templet at R just coincides with the next pitch point T, and the point S with the pitch line, when the next tooth profile may be traced, as dotted at QU. Proceed likewise around the wheel-blank, or the drawing on the drawing-board. Then the templet, being made of thin material, may be turned over and the other sides of the teeth all drawn, and the outlines of the teeth completed, as shown for the one OQU.

A **Radius Rod**, if preferred, may be used instead of the points R and S, the templet being mounted as at X on the rod AX by tacks or screws, and a center pin struck through the radius rod, and at the exact center point A, by means of which the templet may be swung around the wheel and always remaining in the exact tooth-curve position; it being only necessary to stop and hold it at each of the several pitch points, while tracing the tooth curve for the same. The templet may be turned over on the radius rod and the other sides of the teeth drawn.

For Involute Teeth, as in Fig. 180, let A and B represent a pair of pitch lines for which it is desired to draw this form of tooth. Draw a line DCE through the pitch point C and at an angle of about 75 degrees with the line of centers AB. Also draw AD and BE from the centers A and B perpendicular to DE, and circles tangent to DE as shown. To the last-named circles draw a series of tangents JL, GM, KN, IO, as many as required. Then with spacing dividers step from C along on CD to near the tangency D, then backward on the circle an equal number of steps to F, and note that point. Without lifting the dividers, step back past D to near L and out on LJ to J an equal number of steps and note the point J. Then back on JL and on to near M and out on MG to G and note that point. So proceed for as many lines, JL, GM, etc., as the desired frequency of the points F, C, J, G calls for. Then trace the curve through these points, which curve will be an involute to the circle CDM, exactly in theory, and very nearly so by this process of drawing, and to serve for a tooth curve for wheel A. Likewise the involute $HCKI$ is to be drawn to serve for a tooth curve for B.

Then the addendum circles and the root curves are to be drawn in, giving the full tooth profiles, corresponding to *PCO* of Fig. 179, when templets may be made and applied as explained at Fig. 179.

The line *DCE* was said to be drawn at about 75 degrees with *AB*, but this angle is arbitrary, some preferring it at more and some less than 75 degrees. Probably the best criterion to follow as to this

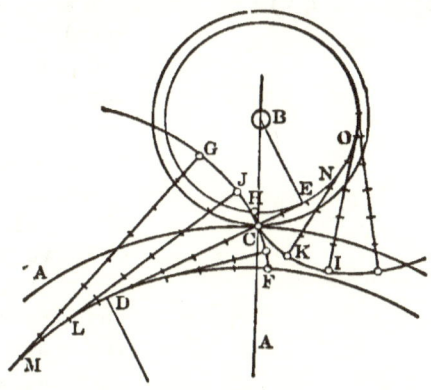

FIG. 180.

angle is the judgment of the designer as to form of teeth a particular angle gives, except that the addendum circle for *A* should not reach beyond the point *E*, nor the addendum circle for *B* reach beyond *D*, on the penalty of interference of teeth; and, at the same time, the normal pitch of the teeth should not exceed the portion of *DE* as intercepted between the addendum circles. The normal pitch is the distance from one tooth to the corresponding point on the next, as measured along on the line *DE*, this line being normal to all tooth curves that intersect it between *D* and *E*.

Approximate Teeth in Practice.

In many cases, a simple circle arc will answer the requirement for a face curve and also flank, especially in gear patterns where the inaccuracy of moulding of the gears in moulding sand enters the account; and for moderate sizes. But it is easy to select a curve that will approximate the epicycloid or the involute closer than the circle will, as, for instance, an ellipse, though a series of ellipses of various sizes would be needed to select from in a particular case for the best results. To avoid this, a parabola or hyperbola may, in a single curve, meet all sizes of teeth, as well as a series of ellipses, but how to locate the curve to best approximate the true tooth without drawing it would be a serious question with a parabola or hyperbola. Further thought suggests a spiral as having advantages, and, finally, the logarithmic spiral, by reason of its peculiar and simple properties, is perceived to be the best adapted of all curves for this purpose. Thus it is well known that a normal to this spiral at any point, a, will be tangent to an equal logarithmic spiral EcF, Fig. 181; and that ac will equal the length of the curve cE, and also equal the radius of curvature of the logarithmic spiral EaG at a. This shows that the two equal copolar logarithmic spirals EaG and EcF are mutual involute and evolute curves. Also, it is known that the triangle Eac is right-angled at E.

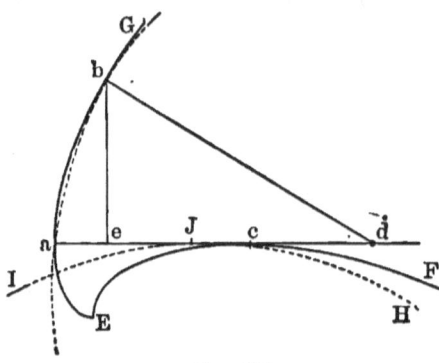

FIG. 181.

It is found by trial on a carefully made drawing that, taking any point d as a center on ac produced, and drawing a circle through a as dotted, the circle will cut the spiral at such point, b, that the length cd, divided by the length be, will be very nearly a constant, at least when the obliquity of the spiral is such that Ec equals $2Ea$, as in the case of the logarithmic spiral, which is found to best fit the epicycloidal tooth curve.

Now draw a pitch line IJH for a gear wheel and tangent to ad at J, so that aJ will equal the half-tooth thickness. Then the epicycloidal tooth curve, suitable for a tooth, so drawn as to pass through a, will also pass very nearly through b, when the latter is taken at the addendum circle; due regard being paid to the pitch diameters and tooth numbers in drawing the epicycloid.

The Template Odontograph has been formed of this spiral with $Ec = 2Ea$, or with the tangent of obliquity equal to 2, and is therefore a curve which very closely approximates the epicycloidal tooth curve. It forms at once a universal and ready-made tooth templet for the draftsman, which has been generally accepted as being a good substitute for the true tooth curve. The instrument is accompanied by tables for use in setting it on the pitch line in the right position for drawing the various kinds of teeth, such as flanks radial, flanks concave, interchangeable sets, involute teeth, etc.*

The instrument, full size, is about four by six inches and is shown in Fig. 182. The curved edge at A is graduated and numbered for a length of 3 inches. To set the instrument in position, "settings" are made out from accompanying tables, as, for example, 2.50. A line ABC is drawn, when the 2.50 is brought to the line at A, while the curved edge at B is brought just tangent to the same line ABC.

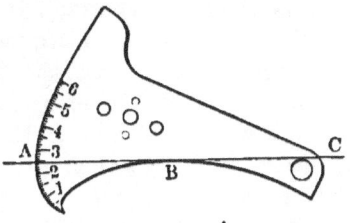

FIG. 182.—Odontograph set at setting number 2.50 on line AC.

To set the instrument for drawing the tooth of Fig. 183, draw the tangent $AHCE$ to the pitch line at C, the middle point of the tooth, also a tangent BD at H, the side of a tooth. Then, with "settings" made out from the tables, place the instrument on the tangent HCE as explained, and, while retaining it there, draw the scriber along the edge from D outward, giving the face curve. Similarly with the instrument set on BD as shown, draw the scriber from D inward, giving the flank curve. The addendum circle and root curves may then be drawn in as previously explained, when the full tooth profile is completed.

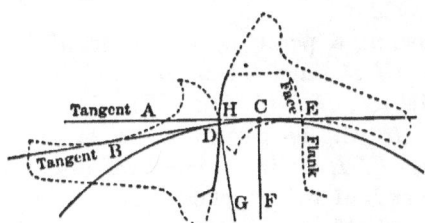

FIG. 183.—How to place the Odontograph.

This *Templet Odontograph*, by means of screw-holes shown, may be mounted on a radius rod and swung around a center pin, for marking all the tooth curves, as explained at Fig. 179. We thus have, in this instrument, a convenient ready-made tooth templet for all cases.

The accuracy of results, as compared with those of other approximate methods, is discussed, and examples are given in the

---

* For a full description of the instrument, formulas for calculating "settings," tables of settings, etc., see *Van Nostrand's Science Series, No. 24*, and *Van Nostrand's Engineering Magazine* for 1876; also a pamphle accompanying the instrument, all found in the instrument stores.

articles cited in reference. Involute teeth, as well as the various cases of rack-and-pinion annular wheels, etc., are drawn by the instrument.

The Willis Odontograph is a well-known instrument for locating the centers of circle arc tooth curves in such a way that a face is of one circle-arc and the flank of another, for teeth approximating those of the epicycloidal form in interchangeable sets, where a wheel of 12 teeth has radial flanks. This instrument is of special interest, as being the first odontograph: originated some sixty years ago by Prof. Robert Willis.

The instrument is founded on principles made clear by Fig. 184, where AC and BC are the radii of the pitch circles, CDH the generating circle to roll on A, to describe a face of a tooth of A and to roll inside of B to describe the flank for B, that is to mate with the face of A. Take the center of CDH on the line AB.

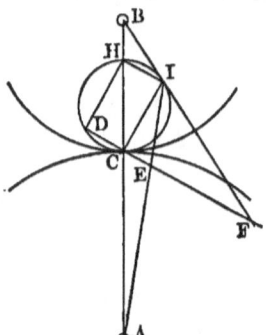

FIG. 184.

Assume a point D on the describing circle CDH, and draw DCF, DH, CI parallel to DH, BIF, and AEI. Then CI and DH will be perpendicular to DCF, because CDH is in a semicircle; and E will be the center of curvature at D of an epicycloid drawn through D by a tracer at D in the circle CDH used as the describing circle, as the latter is rolled on the pitch circle A. Also, F will be the center of curvature at D of a hypocycloid drawn through D by a tracer at D, in the describing circle, as the latter rolls inside the pitch circle B.*

---

* A proof of this is given in *Van Nostrand's Engineering Magazine* for 1878, Vol. XIX, page 313, viz.:

$$CD = HI : CE :: AH : AC.$$

$$\therefore CE = \frac{CD \times AC}{AH}, \text{ and } DE = CD + \frac{CD \times AC}{AH} = CD\left(1 + \frac{AC}{AH}\right),$$

which is the expression for the radius of curvature at D for an epicycloid passing through D as generated by rolling the describing circle CDH on the pitch line A.

In a similar way, the radius of curvature at D of the hypocycloid passing through D, as generated by rolling the describing circle CDH on the pitch line B, is

$$DF = CD\left(1 + \frac{BC}{BH}\right).$$

See Davies and Peck's Mathematical Dictionary, page 222, for formulas for radii of curvature.

Now, as E and F are centers of curvature of the mating epicycloid and hypocycloid tooth curves, it is plain that, by taking the points E and F as centers and describing circle arcs through D, these arcs will approximate the epicycloidal curves and serve as mating tooth curves with corresponding approximation to accuracy.

Without reference to analysis, regard ICF as a triangular templet, swinging about a pin at I. For a movement corresponding to that of the action of face and flank, the edge CF will not depart far from intersection of AB at C. This edge line, CF, would have nearly the same movement if, instead of one side of a triangle, it were jointed at E to a rod AE, since E is on the straight line AI, and would, in either case, move in a circle arc normal to AI. Also, for a like limited movement, if F of the line CF were jointed to a rod BF, the point F would move in nearly the same path, so that motion transmitted from A to B through the medium of rods AE, EF, and FB, jointed at E and B, would maintain nearly constant velocity-ratio for the limited movement considered, as required for the tooth curves.

If now, with E and F as centers, circles be struck through D for a pair of tooth curves for gears A and B, their velocity-ratio would be the same as that for the rods connecting the axes A and B, because the distance between the centers E and F would remain constant in either case. This last consideration holds for whatever point near D in the circle CDH selected, as that through which the circle-arc tooth curves be struck.

Hence, to draw a pair of approximately correct tooth curves as circle arcs, for the pitch lines A and B, Fig. 184, draw the lines AB, CF produced, CI perpendicular to CF, AI, and BIF. Then, with E and F as centers, draw a pair of circle arcs through some point on CF produced. Thus, Fig. 184 is general, but for convenience of millwrights Prof. Willis limits the diagram, the point through which the teeth are drawn being taken a half-pitch from C toward D, and may sometimes be the point D, as when the line CF is drawn, as Prof. Willis takes it, at an angle of 75 degrees with AB.

To obtain a convenient odontograph, Prof. Willis assumes the points I and D to be on a circle $CDHI$ of constant diameter for a given pitch and to equal the radius of a wheel of twelve teeth, for which, D will be a half-pitch from C, and the line DCF at an angle of 75 degrees. This makes all the tooth-circle arcs pass through the point D when DE, the radius of a face, will be the radius of curvature at D of an epicycloid through D generated by rolling the circle CDH on the pitch circle A; and DF the radius of curvature of the hypocycloid in B through D, thus giving for the tooth curves the osculating circles of the epicycloid and hypocycloid at D. A carefully-made drawing will show that the point D for the above limit will be a little above the half-height of the tooth face; while it is probable that a preferable result is obtained if it be a little below, as for the case where a 15-toothed wheel has radial flanks with CD a half-pitch and the line CF at an angle of 78 degrees with AB.

The Willis Odontograph, shown at agF, Fig. 185, locates the

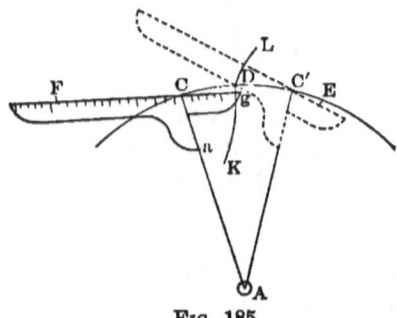

FIG. 185.

centers E and F of Fig. 184 with only the brief diagram of Fig. 185, where CC' is one pitch in the pitch circle, D the middle or half pitch point, AC and AC' radii, and DL and DK a face and flank curve or circle arc respectively, drawn to the centers E and F. These centers are found by placing the odontograph at $C'E$ and at CF, with the leg Ca on the radius, and noting the points E and F by aid of a table from which values, or distances $C'E$ and CF are taken, according to radius of wheel being drawn, and noted in the scale gCF. When the tooth profile KDL is thus drawn, a profile templet may be formed as explained at Fig. 179, and the tooth outlines drawn for the complete wheel.

This odontograph is made in metal or cardboard, the latter being larger and having the necessary tables printed thereon.

A simple odontograph was brought out by Prof. Willis, giving centers of tooth-circle arcs, when the whole profile is formed of one arc, thus approximating involute teeth. This instrument simply gives the centers D and E, Fig. 166, and is of so little help in finding D and E as to be in slight demand.

The Three-Point Odontograph of Geo. B. Grant is so called because its application results in a circle arc struck through three points Cab, Fig. 186, in the actual epicycloidal face curve CD; one point, C, being at the pitch line ICH, a second at the middle of the face at a, and the third at b, where the addendum circle intersects the

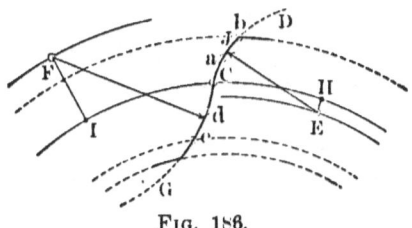

Fig. 186.

epicycloid. The position of the center E, of a circle which will pass through the three points C, a, b, is calculated, and the radius, EJ, called "Rad.," and distance, HE, called "Dis.," are tabulated for a large variety of pitches and sizes of gears.

The flank is treated in like manner for three points C, d, e with the radius Fd and distance FI determined for various sizes and tabulated.

Tables for unit pitch may thus be made out for all varieties of teeth, not only of the epicycloidal order, but involute, conjugate, etc.

The application, in addition to drawing the pitch circle and pitch point, requires simply that "Dis." be taken from the table, multiplied by the pitch and laid off on a radius from the pitch line, as HE, and a circle drawn through the point concentric with the pitch circle. Then the "Rad.," taken from the table, multiplied by the pitch, is to be taken in the dividers, when circular faces Cab may be struck through all the pitch points C from centers taken on the circle through E.

In like manner, the circular flanks are to be struck in from centers on the circle through F by aid of values for "Rad." and "Dis." taken from the table.

The maximum error, or deviation of these circle-arc faces from the true epicycloid, between C and b, is stated to be less than the hundredth of an inch for a pitch of three inches, a quantity hardly worth considering in practice, except in very large, heavy gears with machine-dressed teeth.

The true epicycloidal face for Cab lies outside the circle between C and a, and inside between a and b, and differs from the Willis circular face in that the latter is more nearly parallel to the epicycloid from b to a and leaves a fullness at C. In action, the Willis teeth will thus be more inclined to receive the heavier bearing pressures near C and the lesser near b, or have the working contact between the teeth near the line of centers where the slipping action is least, rather than at a remote point from the line of centers with a greater rate of slip.

By some it is thought advisable to arbitrarily allow the teeth to be somewhat slack near ab, to enable the teeth to perform with comparatively light working pressures here, and heavier at or near C, thus reducing the strains, the friction, and the wear of the teeth. To this end, the radii from the Grant tables may be arbitrarily reduced in length to give a corresponding slackness in the neighborhood of ab.

In Fig. 186, the two points d and e, in the flank, chosen to locate the flank circle arc, provide for a greater depth of flank than ever goes into action on a face, and it seems probable that e should have been chosen not much, if any, below, the working depth near the middle of Ce, in order to the greatest precision of the actual working tooth profile. As a result, the circle arc between C and d makes the tooth fuller than would the actual hypocycloid.

If an "odontograph" is to mean an instrument, it appears that the Grant odontograph is simply any ordinary measuring scale. Thus the pocket rule becomes an odontograph which by analogy with the Willis odontograph is used in determining two distances on the drawing, instead of one angle and one distance.

Circle-Arc Tooth Outlines may be determined directly from the rolled epicycloids or involutes; as, for instance, in Fig. 178, having drawn the addendum circle and supposing it to have cut near G, find by trial a center point that will give a circle arc that approximates most favorably with the curve CFG. Then, noting that center point and radius, describe a circle through the point concentric with the pitch line, and make it the locus of the centers of all tooth-face circle arcs for that wheel, and, using the radius found above, strike in the tooth faces. Likewise proceed for the flanks. Involute profiles require but one center and radius.

Co-ordinating of the Tooth Profiles has been worked out very completely by Prof. J. F. Klein of Lehigh University, whereby the correct tooth profile curve is plotted by aid of co-ordinates taken from a table.

This method would seem to possess special advantages for large teeth where accuracy is required and where the ordinates have a length convenient to lay off. For teeth of one inch pitch or less a magnifying-glass would seem to be a necessity, as well as very fine measuring and drawing devices. Very elaborate tables accompany the method and are embraced in Klein's *Elements of Machine Design*.

These Various Methods of drawing the teeth have each their peculiar advantages, the most advisable for one case not being so for some others. For instance, when very large teeth are to be drawn, with plenty of time for accuracy, the designer may do differently than when the teeth are more moderate in size and more hurried.

All of the above, however, admits of more or less of "hand and eye" operations, in which errors of greater or less magnitude will be unavoidable.

CHAPTER XIV.

MACHINE-MADE TEETH.

Gear teeth, where no "hand and eye" process enters the account, are possible, and in fact are in practical commercial operation to-day, both where the processes conform strictly with theory, and where they only approximate it. Doubtless watch gearing is the best example of the former, the hand and eye operations all being eliminated, as well they may be from the very fact of the minuteness of the teeth.

In Fig. 187 is illustrated what may be termed an Epicycloidal Machine used to cut a tool to the truly epicycloidal shape. Here the tool D to be formed epicycloidal is made fast to the

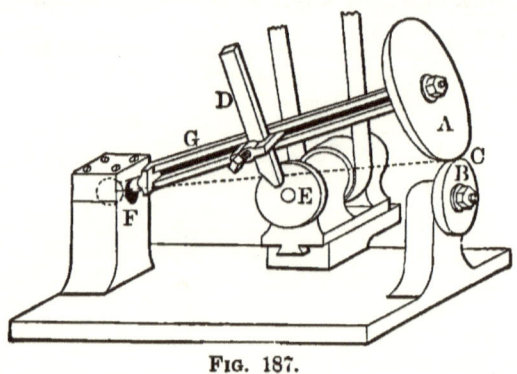

Fig. 187.

shaft G. One end of the shaft is pivoted by universal joint to the post F, while the other end has the disk A made fast to it. This disk rolls upon the stationary disk B. At E is a guiding wheel charged with diamond dust or emery.

The disk A represents the pitch circle of the wheel for the teeth of which the tool D is to be shaped, while the disk B represents the pitch circle of the pinion.

As A rolls upon B, the tool blank D is made to rub against the revolving grinding wheel E. By repeated rolling movements of A upon B, the former being shifted occasionally to bring D to touch the grinder E, the tool will finally become ground so as to touch E slightly throughout the rolling of A upon B, when the tool D will be finished to the true epicycloidal shape.

The pivoting of one end of the shaft or bar G at F has the effect of making A and B the bases of rolling cones with common vertex at F, so that by passing a plane transversely at the tool D we cut the cones in circles proportional to the circles A and B. Thus, the nearer the tool is to F the more minute in effect will the rolling circles become, so that with D at the proper point on G our rolling circles will be of watch-wheel dimensions, while the circles A and B will be large enough to manipulate conveniently.

To prove that we thus obtain the truly epicycloidal form of

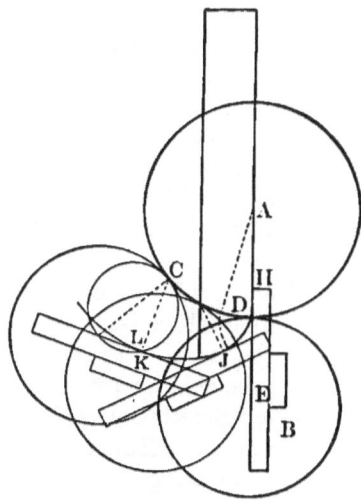

Fig. 188

ground curve on the end of D, we refer to Fig. 188, which represents a section taken at D, through the tool and the cones A and B. Here, however, A and the tool are supposed stationary, with B rolling upon it, and with the grinding disk E also moving around with B. In Fig. 187, the grinder E is placed with its face coincident at the axis of the cone B, as shown in the section, Fig. 188. Then, supposing E to move with B as the latter rolls on A, and noting them in several positions as at JK, etc., we find that the face of E wipes up an epicycloid DJK, the same as if a describing circle, CL, were used with diameter equaling the radius of B, and with a tracing point at L. Thus the end of D has the epicycloidal curve DJ formed upon it, with a continuation from D to H on a radius of A. This line, HDJ, is perceived to be exactly that for a tooth profile for A, answering to the case of flanks radial in both pitch lines B and A. The relative as well as absolute sizes of A and B are seen from the preceding to be entirely arbitrary, so that profiles HDJ may be obtained for a wheel or for a pinion.

The tool D thus obtained is only provisional to the final "fly cutter" for cutting the watch wheels, or the little milling cutter for the pinions; and its application is shown in Fig. 189, where D is used as a planing or turning tool, as the case may be, to form I, the final cutter for cutting the wheel A.

The above process is much simplified by the fact of radial flanks for the teeth. These, for watches, have ample strength and are light-running in action.

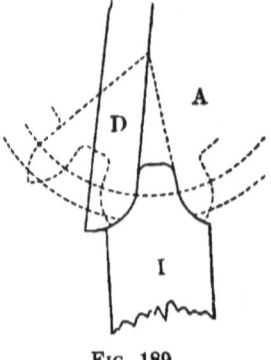

Fig. 189.

The same process may be applied for wheels of any size, and when for ordinary machinery, cylindrical wheels, A and B, may be used instead of the conical ones in Fig. 187, though for cones AF and BF, of considerable length and gradual convergence, the error due to the cones may be reduced to an inappreciable quantity, with results which are practically perfect. Teeth, as above, are independent of all "hand and eye" processes such as locating an irregular curve to draw a tooth face, filing a sheet zinc templet to fit a tooth profile, etc.

The teeth as produced above, all having radial flanks, though entirely suited to such special sets of gearing as used in watches, clocks, and some heavier machinery, where interchangeability is not sought, are not adapted for outfits of simple sets of cutters for general commercial purposes, either in the tool store or in the machine shop. For this it is generally admitted that there are but two systems of gearing and cutters in use, viz:

The Involute and the Epicycloidal Cutters.

These, for cutting interchangeable gearing of its own kind are in extended use: the first in accordance with principles explained in Fig. 166, and the second following the principles of Fig. 153.

In either of these systems, cutters are made in series, each of which will cut several wheels of differing tooth numbers, one of which gears will be right, the others slightly inaccurate, but none out of the theoretic condition to an extent to produce appreciably bad-working wheels.

The first system of cutters of this kind for cutting gear wheels were put out by the Brown & Sharpe Mfg. Co., and were approximately involute in form of tooth cut, each pitch embracing eight cutters to cut gears of from twelve teeth to a rack, and interchangeable—one cutter cutting gears of twelve and thirteen teeth, the next fourteen and sixteen, the next seventeen to twenty, the next twenty-one to twenty-five, the next twenty-six to thirty-four, the next thirty-five to fifty-four, the next fifty-five to one hundred and thirty-four, the last one hundred and thirty-four to a rack.

The second system of cutters for epicycloidal teeth of wheels in interchangeable series are also made by the Brown & Sharpe Mfg. Co., but by machinery brought out by the Pratt & Whitney Co. This system embraces twenty-four cutters for each pitch, cutting from twelve teeth to a rack, a wheel of fifteen teeth having radial flanks. For a full description of the machinery for making these cutters, with no "hand and eye" operation, see MacCord's *Mechanical Movements*, page 178.

There are two machines, one of which is called the Epicycloidal Milling Engine, which mills or cuts out all the tooth templets required in the series, of magnified size; and the other is the Pantagraphic Cutter Milling Engine, which applies the above templets in making the gear-tooth cutters reduced to any desired size or pitch. Other templets may also be used on this last-named engine.

The number of cutters in a series is arbitrary and depends upon the inaccuracy allowed to be admissible. For large cutters, as for large teeth, it would seem that the number of cutters in a series should be relatively great, that the admitted errors of cut teeth may not exceed a certain arbitrary value. A formula for determining the so-called "equidistant series" of cutters is given by Geo. B. Grant in *Teeth of Gear Wheels*, page 20. Experience in the management of gearing would lead one to adopt a tooth numbering in the above series of cutters such that the cutter that cuts gears from 50 to 60 teeth, for instance, will make the 50-toothed wheel right and the others slack on the point, rather than the contrary, that the teeth, in action, may have the severest pressure of contact near the line of centers rather than remote from it, as explained at Fig. 186.

Prof. Edward Sang's Theory of the Conjugating of the Teeth of Gear Wheels in interchangeable series has been applied in two different and patented gear-cutting engines, the first by Ambrose Swasey of Cleveland, Ohio, Pat. No. 327,037; and the other by Geo. B. Grant of Lexington, Mass., Pat. No. 405,030.

In the first, or the Swasey Engine, a split multiple cutter is employed to cut the gear teeth directly; while in the second a solid worm or screw hob is used in cutting the gear. In each of these one and the same cutter cuts all gears of a given pitch, regardless of size of wheel, and giving theoretically correct teeth for all without the intervention of hand and eye operations.

Let Fig. 190 represent Sang's principle in the case of Fig. 153, as applied to the rack where the describing circles for face and flank are one and the same, carrying the tracer D to trace the face CD, and the tracer E to trace the flank CE of the rack tooth as the describing circle rolls along the straight pitch line. These curves for face and flank are both cycloids and identical in form.

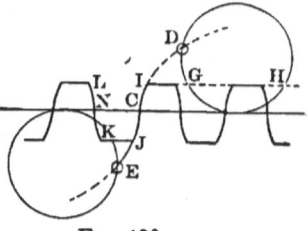

Fig. 190.

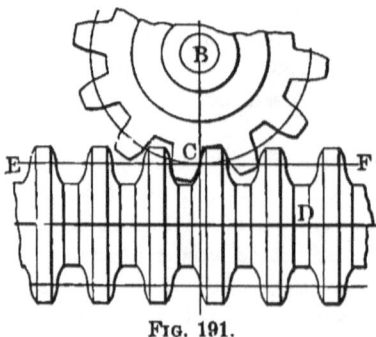

Fig. 191.

Now, as any wheel of the interchangeable series of Fig. 153, of the same pitch and describing circle, will work correctly with this rack, it is plain that if a multiple cutter were made with cutting teeth of the same longitudinal section as Fig. 190, that cutter would cut any gear of the series referred to, if, while cutting, the cutter could be moved relatively to the gear, as A is relative to B in Figs. 123 and 124; or as shown in Fig. 191, where D is the multiple cutter, and

B the wheel being cut while it moves along as if its pitch line were rolling on the pitch line EF from E to F, the cutter at the same time revolving and cutting.

This would seem to require a cutter as long as the circumference of the gear. To avoid this, Mr. Swasey splits the cutter, D, into halves, so that the idle half can move back a pitch while the other half is cutting and moving with the periphery of the wheel, the latter being kept in steady revolving motion with its axis stationary. In this way, the cutter, D, may be so short as to reach from the intersection of its addendum line with that of the describing circle on one side over to like intersection on the other side, or from G to H, Fig. 190.

In cutting a gear, B, the latter is kept steadily revolving, and also the cutter D, in exact relation by gearing, the cutter making as many revolutions to one of the gear blank to be cut as there are to be teeth in that wheel. Then with B and D in motion in the cutting engine, a slow feed is given to the cutter, D, toward the wheel blank, B, cutting all the teeth together by one continuous action. When the gear is partly cut, all the teeth are cut to the same extent, and all are finally finished at the same time.

To form the cutter D, Fig. 191, a tool is required which, on its cutting end has the exact shape of a space in the rack, or of $ICJKNL$, Fig. 190. This tool may best be formed by some such device as the Pratt & Whitney Pantagraphic Cutter Milling Engine mentioned above, if it is to have the cycloidal form without hand and eye processes.

But in the Sang's theory we are not confined to the cycloidal form of curves ICJ and KNL; for a brief consideration will show that any arbitrary curve IC may be adopted and copied at CJ by retaining the point C in common and swinging I around to J in the plane of the paper, and then the whole curve ICJ turned over and copied at KNL. Thus, for the pantagraphic milling engine, a circle may be cut in a lathe for the parts IC and CJ of the templet.

Again, and for the simplest case under Sang's theory, IC may be a straight line, which should be located at a certain angle of IJ with NC, usually about $14\frac{1}{2}$ degrees. Also the same for KL.

This last-named case gives us the well-known series of interchangeable wheels with involute teeth.

This multiple cutter, thus made in two parts, is, in effect, turned in a lathe, by cutting one space groove after another with the tool formed as above explained, the top face of the tool being held in the meridian plane of *D* while cutting it.

In the Grant Cutting Engine the splitting of the cutter *D*, thus complicating and weakening it, is avoided by the use of a solid worm cutter or hob, the latter making one revolution while the gear turns one pitch. In cutting a gear, the motions of the cutter and gear bank are continuous and in definite relation till the gear is finished, similarly as in the Swasey machine, and this machine would seem to have the preference, unless it is found that the hob cutter is so complex in form of worm threads as to be unduly difficult to make.

Investigating this, we find that hobs for cutting *epicycloidal teeth* must have the axes of the hob inclined to the plane of the gear being cut by an angle equal that between a plane normal to the axis of the hob and the tangent to the worm thread of hob, at the pitch line. That is, the element of thread of hob, at the pitch line, must be perpendicular to the plane of the wheel. This inclination can easily be found from a triangle where one side is the circumference of the hob at the pitch line, another side, perpendicular to the first, the pitch; in which triangle the smaller angle is to be taken for the inclination.

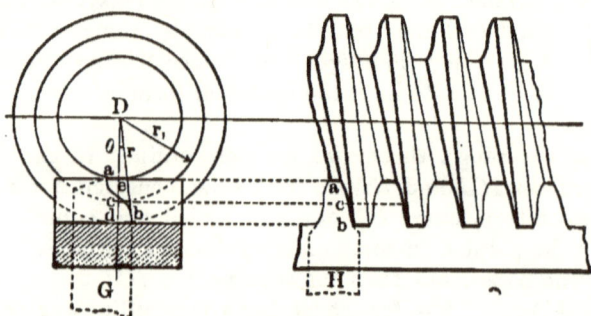

Fig. 192.

Even with this inclination of the thread of the hob (speaking of it as before the teeth are cut in it and when it is simply a screw), it will not be tangent to the cut teeth of a gear, or of a rack, Fig. 190. cut by it, except in an S-shaped curve, *acb*, Fig. 192, which is more pronounced as the radius *r* is shorter and not vanishing until *r* is infinite.*

---

* The equation of this curve *acb* is $r \sin \theta = r \dfrac{-AB + \sqrt{1 + B^2 - A^2}}{1 + B^2}$, in

To give an idea of the intensity of the S-curve for epicycloidal teeth, take $r_1 = 2''$, $p = 0.5''$. Then db, Fig. 192, is $0.158''$; and for the ten equal divisions from d to a, the remaining ordinates are $0.149''$, $0.126''$, $0.104''$, 0.075, $.0''$, $-0.078''$, $-0.107''$, $-0.129''$, $-0.147''$, and $-0.162''$, the extreme ordinates being the greater.

The tool which cuts the worm, as in a screw-cutting lathe, should have this shape on its top side at the cutting end, as shown dotted at G and H, in case the tool is to be used as in cutting other threads. But a preferable way is to make its top flat and straight and of the true epicycloidal shape in plan as at H, and, after cutting in the ordinary way to the right depth for the hob thread, then, without varying the depth of cut, make several passes for cuts with the tool at varied heights, raising and lowering on a line which is tangent to the hob thread at the pitch line, and for a range covering the ordinates db and ea. A safe way would be to work the tool at different heights thus as long as it will take a cut. This supposes that the tool for cutting the hob thread has the correct shape as at $ICJKNL$, Fig. 190, and formed as already explained.

But in cases where the tooth profiles are not normal to the line CN, Fig. 190, the cutter hob will not need inclining so much, and, in some cases, not at all, as for the involute series.

The hob for involute toothed wheels in interchangeable series for use on the Grant Cutting Engine is to be made in a similar way if its axis is to be inclined as before, so that at the finishing of cut the tangent to the pitch-line element of the hob thread is normal to the plane of the wheel being cut. But for this case of involute gearing it is not necessary to incline the axis, as it may be kept parallel to the face of the wheel being cut, if the hob is formed with due regard to this, viz.: that, in cutting the hob threads, as in a threading lathe, the tool be held at varying heights as before, while cutting, except that now it is to be raised and lowered on a vertical line instead of a tangent to the pitch-line element, and to

---

which, for epicycloidal series, $A = \dfrac{r_1}{r}$ and $B = \dfrac{2r_1}{p}\sqrt{\dfrac{r - r_1}{7.5p - r + r_1}}$, in which last, p is the diametral pitch.

In involute series with axis of hob inclined as above stated, $A = \dfrac{r_1}{r}$, and $B = \dfrac{2r_1}{p} \cdot \tan 14\tfrac{1}{2}° = 0.52 \dfrac{r_1}{p}$. In constructing these curves, r is to be laid off from D on Dc; and $r \sin \theta$ is to be laid off from Dc on circle arcs struck from D.

continue the raising or lowering and repeated cuts as long as it will take cuts. In this case, however, the tool may be placed at the height, first above and afterwards below the axes of the hob, in cutting the thread equal the diametral pitch divided by twice the tangent of the angle of inclination of the edges of the tool, or of the sides of a tooth in the rack, which inclination is usually about $14\frac{1}{2}$ degrees. Hence the height above or below is about twice the diametral pitch.

In all cases above, the thread tool for cutting the hob threads is to be correct in shape and dimensions as answering to the outline $ICJKNL$, Fig. 190, except that it should be somewhat in excess of length so as to cut the hob deep enough to have clearance at the top of the teeth, and with due regard to side clearance between resulting teeth of wheel.

It appears that the hob threading tool must have the correct shape in any event according to the system, and of the right thickness for a tooth; but that the threads of the hob are not of correct tooth profile shape in any case. For involute teeth these threads will not be straight on the meridian intersections. Hence teeth of spur wheels cut in this way in the Grant Cutting Engine with the ordinary worm-wheel hobs will not be correct in profile of tooth.

In these gear-cutting engines of the Swasey or Grant order the spacing of the teeth may be expected to be unusually even and exact, since in the cutting the hob acts upon several teeth at the same time, and the wheel being cut may be driven around by worm and wheel. The latter may be made in two half wheels joined on a plane transverse to the axes and through the middle of the teeth. In cutting this duplex worm wheel, when partly cut it may have one half loosened and shifted on the other half the space of a few teeth, made fast and some further cutting done, then shifted again and again, till the teeth in the two halves will all fit exactly for any way of combining the halves. Then when in use and wear the part may be occasionally shifted. In this way, with these cutting machines, it seems certain that the spacing as well as the tooth outline of cut gears may be made marvels of exactness.

Tooth Planing or Dressing Machines.

These machines, like the Gleason's, are in use in some machine-shops, where the teeth of large heavy cast gears are tool dressed, as in a shaping machine, the tool being guided by a templet which may have been formed by hand, thus admitting hand work in part.

George H. Corliss appears to have been the first to do this, his first work dating back to the 40's or 50's: the heavy fly-wheel gear on the *Centennial Corliss Engine*, central in Machinery Hall, being one of the more notable examples of tool-dressed spur wheels. Large bevel wheels, connecting the main shafting with the above engine were also tool dressed on a bevel " gear planing " machine exhibited by Mr. Corliss.

Hugo Bilgram exhibited remarkably smooth-running bevel-gears at the World's Fair of 1893, the teeth of which were dressed out in a similar way, as an example of commercial work by him.

In these tooth-planing machines for bevel gears the point of the cutting tool is made to move on a slide in a line joining the vertex of the cone of the gear and the point of contact of the guide finger with the guiding-tooth templet, this tool and finger being mounted, in effect, on a universally swinging arm, pivoted at the cone vertex. Thus the elements of the finished tooth are made to converge to the apex of the conic gear. Large or small gears can be tooth dressed on the same machine. One considerable advantage here over the spur-gear tooth-dressing machine is that the tooth templet may be made of magnified size, so that bevel wheels, compared with spur wheels thus tooth-dressed, may be regarded as more nearly perfect, while the contrary is true of wheels cut in the ordinary way with revolving cutters.

Stepped and Spiral Spur Gearings.

In some of the finer light-running machinery at high speed the plain spur gearing is quite likely to make a humming sound as the teeth engage in succession, each contact giving a slight click.

This is avoided in a measure by stepped gears, the wheel being made up of several thin ones made fast together and arranged in steps so as to divide the pitch into as many parts as there are thin parts in the wheel.

Another way is to use the spiral gearing of Hooke, in which each tooth is on a spiral slant, to such extent that one pair of teeth engage before the preceding pair quits engagement.

This causes endlong pressure on the axes, objectionable in heavy working wheels, but which is often obviated by making each tooth of equal portions of right and left handed spirals.

CHAPTER XV.

SECOND: AXES MEETING.

BEVEL GEARING.

The teeth of these wheels may be made of any of the forms explained under the case of axes parallel. The pitch surfaces being cones, with vertices at a common point, the describing curves, by analogy with cylindric gears, are also cones, with vertices in common with the pitch cones.

Theoretically Correct Solution.

In practice the bases of the cones may be made spherical, with centers at the cone vertices, and the describing cones may be realized in concave templets fitting these spherical surfaces, as explained in Fig. 138. In this way, the correct tooth curves may be laid out on the gear blanks, the number and kinds of templets required being explained in Figs. 111 and 112, though here all made concave and fitting the spherical base of the cone.

In these wheels, no circular rolling cone can describe precisely radial flanks, and to approximate them the describing circular cone must be made a trifle larger than to span the distance on the sphere from the pitch line to the pole or axis to the wheel.

Approximate Solution.

But a much simpler way, and that usually followed in practice, is Tredgold's approximate construction, which, though slightly inaccurate in theory, is appreciably exact in practice. In Fig. 193, take A and B for the wheels with pitch surfaces in contact at CG, they being of conical form, with the cone vertices at a common point in O, the axes being OAD and OBE. The length CG is arbitrary, and also the angle of intersection of the axes at O. In practice the latter is usually a right angle.

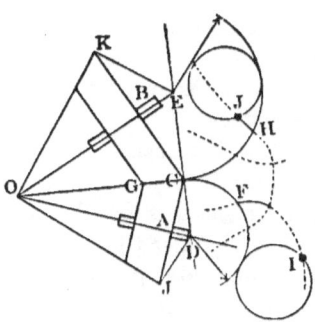

Fig. 193.

Tredgold draws a line, DCE, perpendicular to the line or element of contact CO of the pitch cones, when, if CD be revolved about the axis AO, a conic surface $ACDJ$ will be generated, which is normal to the pitch cone, $ACOJ$. Likewise will the line CE generate the cone $BCEK$, normal to the pitch cone $BCOK$. Developing these cones from the line ED, we obtain the circles of development DCF and ECH, with centers at D and E.

Upon these circles the teeth are laid out as if they were the pitch lines of a pair of gears, any system of teeth being selected as preferred. For epicycloidal teeth with concave flanks, the describing circle carrying the tracer J has a diameter which is less than CE, and in the usual way is rolled inside of CH and outside of CF, to generate a pair of mating tooth curves, a face and flank. Another describing circle is rolled on the other sides of the circles CF and CH, thus completing the tooth profiles as dotted.

The addendum circles, root circles, clearing curves, etc., may be drawn in, as in spur gearing, when the conic surfaces CDF and CEH may be re-developed or returned to the cones $CDAJ$ and $CEBK$, taking the tooth drawings with them, giving us the laid-out teeth on the conic blanks, as in Fig. 194.

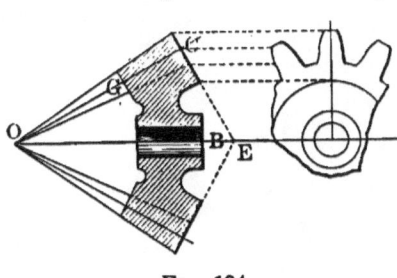

Fig. 194.

In preparing the blanks, they should be left larger to include the addenda and other portions of the finished wheel as shown in Fig. 194, unless it is preferred to dress the flank to the root surface and add the teeth thereto as often done.

The tooth-surface elements should all vanish at O, and a convenient as well as sure way to give the teeth the right directions in the finished wooden gear patterns, for instance, is to fix a fine line at the point O, which may be drawn to the tooth outline at C to determine when the tooth is dressed to the proper lines, element by element.

Bevel wheels are not practicable in interchangeable series, because, if one pair have axes at right angles, the substitution for one of these of another correct working wheel of larger or smaller radius changes the angle between the axes to something other than the usual 90 degrees. Therefore the teeth, and both wheels entire, must be made in pairs and of shapes to suit, regardless of interchangeability.

Spiral Bevel-wheel Teeth.

These are possible, but not common, because difficult to make, except by special machinery, which probably does not exist. Stepped teeth would be more readily made.

Bevel-wheel teeth, carefully planed to shape by templet in a gear-planing machine, will work fairly well. See Tooth Planing, under Axes Parallel.

CHAPTER XVI.

THIRD : AXES CROSSING WITHOUT MEETING.

SKEW-BEVEL GEAR WHEELS.   APPROXIMATE CONSTRUCTION.

As stated in connection with non-circular skew bevels, a method of laying out theoretically correct and practicable teeth is not known, except for gears taken at or near the common perpendicular between the axes, or near the gorge circles.

But teeth which approximate the epicycloidal form may be generated by employing generating hyperboloids to roll upon the pitch hyperboloids inside and outside, in a manner analogous to the use of describing cones in common bevel wheels, and as shown in Fig. 138. These approximate tooth curves and surfaces become more and more inaccurate as the angle between the axes increases toward 90 degrees, at which limit considerable interference occurs, requiring the teeth to be "doctored" by arbitrarily dressing off certain tooth faces or mating flanks to an appreciable extent, to get the wheels to work with acceptable smoothness.

The patterns for cast gears may thus be executed to better advantage, probably, than in any other way, even not excepting Prof. MacCord's construction, based partly on Olivier's theory of involutes for one tooth surface, assuming another, and determining the rest by difficult conjugating.

To construct these approximate skew bevels, let $PRLNSQ$, in Fig. 195, represent the blank of a skew-bevel wheel, for which the hyperboloid of revolution $ITUVKM$ is the pitch surface, determined as in Fig. 10, extended from IM to and past the gorge circle, TK, reaching UV at a distance $OZ = OA$ beyond the gorge circle. The addendum and dedendum surfaces PQ and RW are drawn in throughout, being hyperboloids of revolution, because a top center line of an extended tooth, as well as a root line, being straight, would, in revolving, sweep up hyperboloids of revolution, for the same reason as would the element CO of the pitch surface. At JK these three lines are shown as parallel to the axis in projection, which determines the height, ab, of a tooth at the gorge circle.

Then the contour lines of the addendum surface can be drawn in as hyperbolas, $NSbd$ and $LYae$. The lines NL and SY should be extended normal to MK, to the bottom of the web upon which the teeth are mounted. With this much drawn, the skew-bevel blank can be turned up, arms and hub being assigned at will.

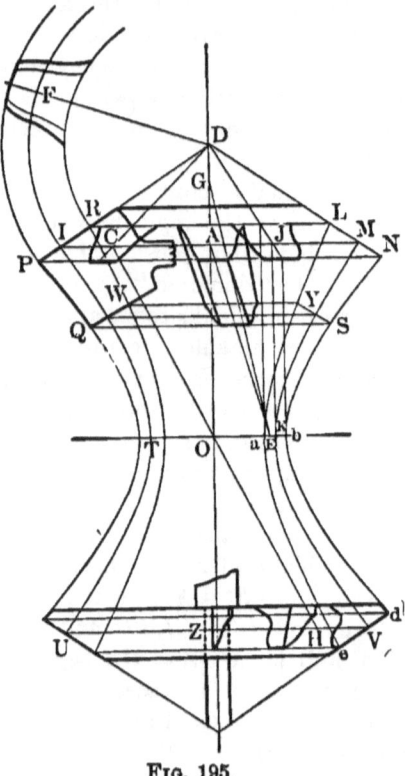

FIG. 195.

Tredgold's method is adopted, as in Fig. 193, of developing in IF the normal cone (normal to the pitch surface) upon which to lay out the teeth at F. The epicycloidal or involute form of tooth may be chosen here, as well as in ordinary bevel gearing, but this example will be carried out in the approximate epicycloidal form, with a view to testing the theory of the generating hyperboloids.

The rolling circle applied here gives a close approximation to the normal section of a tooth at I, as would be developed by the describing hyperboloid above explained; so the full drawing of the skew tooth will be first given, and its errors afterwards sought out.

Instead of the normal section of tooth, we require an oblique section, as shown at A, formed by the intersection with the tooth of the surface of revolution $NLRP$. To obtain the excess of width of this oblique over the normal section, revolve the line CO (same as CO, Fig. 10) about the axis, till C falls at A and O at E. Then revolve the center line EA of the tooth about E, till A falls at G, where EG is parallel to the plane of the paper, when EG appears in its true length. A convenient way to find G is to make EG equal in length to CO, G being on the line AD. Then the angle OGE is the angle of obliquity of the normal section of tooth with the oblique section at A. Therefore the width of the oblique compared with the normal section of tooth at F is in the ratio of GE to GO. This may be laid off at several points in the height and the oblique section determined. If preferred to do it now, the faces and flanks may be arbitrarily doctored from the pitch line, each way, to prevent interference, as explained above in connection with Fig. 196.

With the oblique section determined, as in the outside lines at F, we may cut a templet in thin material and use it on ICM, as explained in Fig. 179, to delineate all the teeth on the pitch line IM.

Then, in cutting the teeth to shape, the oblique direction AE is probably best arrived at by carefully cutting a thin piece UV to all the tooth outlines, beveled back in excess from the tooth profiles, and mounting it on the axis AOZ so that OZ equals OA, and in such position that the line CO from the center of a tooth at C will strike the center of an outline at H. Then a thread may be drawn at any time, in dressing a tooth, from any point on H to the like point on C, and each element of C thus dressed the whole length of the tooth to fit the line. This process, repeated for all the teeth, completes them unless when the mating wheel is likewise thus far completed and the pair be placed in running relation it be found that interference exists. If so, the teeth are to be dressed off at will, till interference is relieved, which completes the wheel.

These wheels may thus be made of any desired length from C toward H. To give an idea of the amount of interference of the epicycloidal teeth as above, Fig. 196 is introduced, which was drawn with great care and labor from a model like that described by Geo. B. Grant in the *American Machinist* for Sept. 5, 1889, Fig. 3. In the present example the axes are at right angles, at a distance of 3.7 inches apart at the common perpendicular, the wheels being 11 and

6.8 inches in diameter respectively, and with the velocity-ratio of 3 to 2. These tooth-curves were traced with the true rolling hyperboloids.

The figure is a transverse section, taken on a perpendicular to the element of contact, *CO*, Fig. 195, and at a distance of 5.8 inches

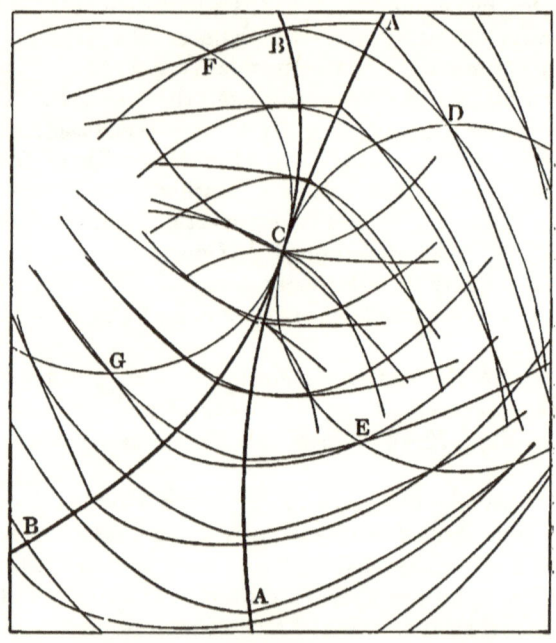

Fig. 196.

from *O*. It cuts through the two pitch hyperboloids in curves of intersection *AA* and *BB*, and the describing hyperboloids in the curves *DCE* and *FCG*, which curves of course are all ellipses. Also it cuts all the mating tooth-curves at three quarters of an inch apart on the pitch line.

The describing hyberboloids are both of one size, and half as large as the smaller wheel, so that by theory the flanks in the smaller wheel should be radial. The figure shows them to be very nearly so.

This is a somewhat extreme case of proportions, and we might expect high per cents of interference. The figure shows almost no interference on the left-hand side of the pitch lines up to a pitch of over two inches; while on the right-hand side, for a pitch of ¾ inch, 1½ inch, and 2¼ inch, we find the faces and flanks cutting into each other to a normal depth of 5, 6, and 10 **hundredths of an inch** respectively, or about four per cent. of the pitch. For a pitch of eight inches the interference depth reaches 15 hundredths of an inch, or about two per cent. of the pitch.

Also, the figure shows that below C the faces of the smaller wheel must be trimmed off toward the point and on the lower side for pitches over about 1.5 inches, while the flanks of teeth of the larger wheel above C seem to need dressing out on the lower sides. The figure may thus aid in determining how to doctor the teeth in a practical case, for best results.

One noticeable feature of these tooth curves is that, if they interfere, they intersect at the surfaces of the describing hyperboloids, in the central positions shown, viz.: on the lines ECD and FCG,—a fact also pointed out by Grant in 1889 in the few curves shown.

Combined punching and shearing tools have had skew gears some 20″ diameter.

Exact Construction.

The exact construction of these teeth has been best treated by Grant in the *American Machinist* and also in his valuable work on *Teeth of Gears*, where the theory of Olivier and treatment by Herrmann are discussed.

Olivier appears to have originated involute tooth surfaces, which he calls spiraloids, that may be generated by the revolution of a straight line about a cylinder, the line being maintained at a constant angle with the cylinder and prevented from lateral slip. The same result is accomplished by rolling a plane around a cylinder without slipping, on which plane a line is drawn obliquely. This line will sweep up, in the space about the cylinder, the spiraloid of Olivier, which spiraloid surfaces are proposed for skew-bevel teeth.

Any point in the above line will evidently describe an involute about the cylinder, and all points of the line together, will describe an involute spiral or spiraloid.

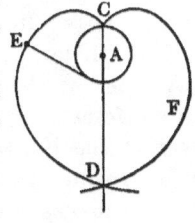
Fig. 197.

In this description it is immaterial to the result if the plane slips on the cylinder in the direction of the line, that is, if the line be only allowed to slide on itself, not laterally, and the angle between the line and cylinder be preserved constant.

Fig. 197 shows a section at right angles to the spiraloid as swept up by one line on the plane, *CED* being cut by the line at one side of that point of the line which comes to touch the cylinder *A* at *C*, and *CFD* cut by the line at the other side of the same point.

THE OLIVIER SPIRALOID.

In a general view this spiraloid with section, Fig. 197, appears like a twisted bar in which the cylinder *AC* is straight, the depression *C* being like a spiral or helical crease along the length, and the edge *D* like a spiral ridge. According to what is stated above, the curves of cross-section *CED* and *CFD*, Fig. 197, are the ordinary involutes to the circle *AC*.

In Fig. 198 two views are presented of a multiple spiraloid of six ridges and creases, the intersection of each ridge by a normal plane giving two equal limited involutes, as *CD* and *ED*. Extended involutes give Fig. 203.

A straight line, *FGH*, will touch one ridge from *F* to *G*, while from *G* to *H* it lies in contact with the under side of the next ridge above *FG*. Therefore these ridges are enveloped by surfaces which have involute elements in normal section and right-lined elements in lines tangent to the cylinder *ACEJ*, as in the example of the line *FGH*.

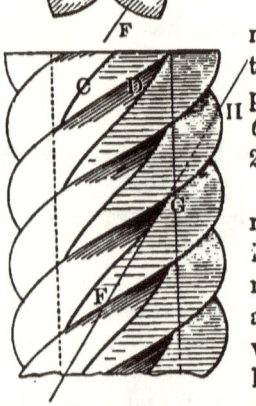

FIG. 198.

Interchangeability of the Olivier Spiraloids.

Now if we remove the alternate ridges or threads, and cut clearance grooves into the cylinder to sufficient depth, as shown by the dotted lines *IJK*, this multiple spiraloid will work, according to Olivier, as a skew-bevel gear with another like it, or with any other of the same normal pitch, regardless of the axial or the circumferential pitch of the helical threads or teeth. This is true even at the limit where the helical ridges become parallel to the axis, provided that the normal pitch is still the same and that the normal planes cut the ridges in involute lines of section, as at *CD*, Fig. 198.

This is equivalent to saying that a numerous set of these spiraloids of the same normal pitch will work together interchangeably. Herrmann pronounces against this.

Interference of Olivier Teeth.

But Olivier's claims can be shown to be true for these gears as cut off at the intersection of the axes, as shown in Fig. 199, when so cut that no contacts of teeth occur outside of the angle AOB, because interference of teeth takes place outside these limits, of the same kind as found to occur between involute teeth of cylindric

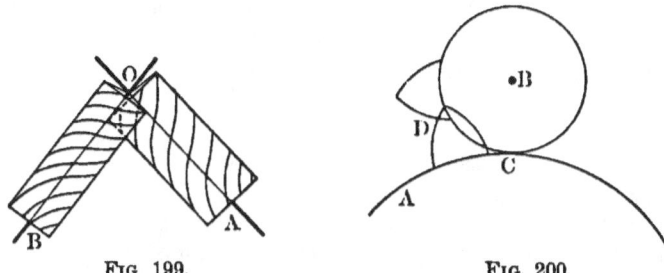

Fig. 199.   Fig. 200.

gears with involutes drawn from the pitch circles instead of base circles within. An example is shown in Fig. 200 at D.

By undercutting the teeth in the vicinity of O, Fig. 199, and to some extent up the face CD, Fig. 198, with plenty of bottom clearance, the teeth will work when the gears extend both ways past O, Fig. 199, if the addendum is not excessive, as has been shown by Oscar Beal in good working gears of the kind in metal.

Nature of Contact of Olivier's Teeth.

Herrmann claims that these gears, if working at all, will have tooth contacts only at points instead of lines, but it can be shown that the contacts will be in lines like FG, Fig. 198.

Olivier holds that the gears as cut off at O, Fig. 199, will drive only in one direction, and that to drive both ways they must be extended beyond O. This is proved by the Beal gears to be true, as well as the necessity for clearance for thick addenda, as above explained.

Thus we have very satisfactory and perfect working skew-bevel gears of the kind shown in Fig. 198, with straight line contacts between good shaped teeth, though the gears extended both ways have not contacts far from the gorge circles, but are capable of driving either way. These gears appear, however, like cylindric skew gears, but are in reality skew bevel, as taken at the gorge of the hyperbo-

184   PRINCIPLES OF MECHANISM.

loids, though with no fixed relation between the gorge radii, and angles between axes and line of contact.

If the gears are at right angles the working straight line contacts will all be within the rectangle $abcO$ for driving one way, and in the rectangle $defO$ for driving the opposite way, while the appreciable interference will be outside the dotted curves g and h, as shown near O. The rectangles are determined by the intersections of the addendum surfaces with the common tangent plane between the two cylinders A and B, which cylinders form the bases of the involute teeth. The greatest working length of gear A is therefore ce, and of B it is bf.

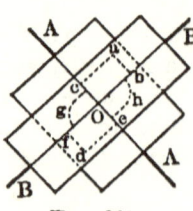

FIG. 201.

RESULTS FROM AN EXAMPLE.

To better fix the ideas, a particular example is referred to of the above cylindric skew-bevel screw-gears, in which the gorge-circle cylinder pitch surfaces extended indefinitely each way, as shown in Fig. 202, are 5.1″ and 2.2″ in diameter, with axes at right angles. For an addendum of 0.32″ the interference was found inconsiderable, but for thicker addenda appreciable interference would begin at about the dotted lines g and h, Fig. 201, when gO and hO equal about 0.8″. Then, in case of the thicker addenda, with clearance for interference cut from C up toward D, and E up toward D, Fig. 198, on the face the necessary amount, it is found that there will be no contacts at or about O, Fig. 201, for a distance from O of about the same value as the distance from the dotted curves g and h over to where the interference for the actual addenda ceases, and probably within some such shaped outline as the curves g and h. These quantities are admitted to be only approximate.

We may Demonstrate the Principle respecting the above exact construction of Olivier by use of Fig. 202—in some respects the same as one given by Grant. Here, for the pitch surfaces of the proposed skew gears, we take AD and BE, the cylinders of the gorge circles, touching at O, between which cylinders in common tangency is a plane GG, upon which are parallel lines, the normal pitch of teeth apart. Now suppose the plane to be moved along with its parallel lines maintained parallel to a fixed line, and that the cylinders AD and BE are revolved by the plane in such a way as to allow no lateral slipping of the parallel lines of G on the cylinders, while at the same time admitting endlong sliding of the lines in their own tracks on the cylinders to any extent. It is plain that if the lines could leave an impression of themselves on the cylinders the latter would appear like perfect screws, having a number of linear threads.

Now suppose that the cylinders had been covered with some easily cut material during this rolling, and that the material within the surfaces cut by the lines on G remained attached to the cylinders. It is plain that the cylinders with these thread ridges thus

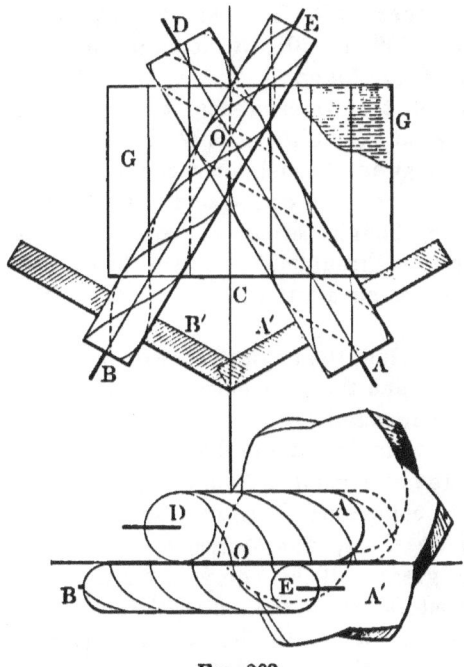

Fig. 202.

formed and attached would be like that of Fig. 198, the surfaces of the threads being of involute form in normal actions and with straight-lined elements in all lines tangent to the cylinders, as for FGH, Fig. 198. Also, it is easily seen that these surfaces are tangent to each other within the angles AOB and DOE above and below O, and that these lines of tangency will all be in the plane G, because this plane is normal to all the cut surfaces. Theoretically, then, these threads, or spiral ridges, will have perfect right-lined contacts within the angles AOB and DOE to any extent from O, and exactly the form of contacts required for gear teeth.

In the above, no particular angle has been assumed between the cylinders, nor between either one of them and the cutting lines on the plane G. It appears, then, that the cylinders may have any possible angular relation with each other, or with the lines on the plane G; and that if the lines be assumed as running from O into the angle AOE instead of the angle AOB the contacts will all be within the angles AOE and BOD, so that the case is perfectly general.

In all cases there will be some degree of interference, as explained at Figs. 199 and 201, in the angles opposite those within which the contacts are found.

On account of this interference it is advisable in practical gearing of this kind extending past O that the addenda be comparatively light—say in the neighborhood of a fifth of the mean gorge radius—and the teeth correspondingly small.

In these gears, though perhaps called skew bevels, if the teeth extend to sharp tops throughout they will be cylindrical screws, as in Fig. 198, there being alternating ridges or teeth, and spaces, as at D and I.

In practical gearing, the angles between the lines of G and axes A and B must be such that the normal pitch will divide the pitch cylinders without remainders.

Gears of this kind, instead of being taken at the gorge circles O, may be selected at a distance from the gorge circles, as at A' and B', Fig. 202, and comparatively short. But if the teeth go to sharp tops they will be short, many-threaded, cylindric screws. The teeth of these gears will be comparatively flat and practically useless if selected far from the gorge, though they will work perfectly in theory. The gear A', Fig. 202, is shown in Fig. 203 with skew omitted, where AJ is the cylinder AD of Fig. 202, and A' the gear.

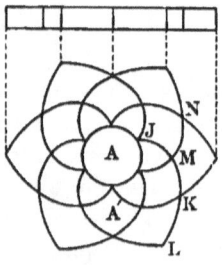

Fig. 203.

The involutes, as extended, intersect at various points M, N, etc.; and the larger gears, like A', require no bottoming clearance, such as needs to be provided for gorge-circle gears, as at IJK, Fig. 198.

It finally appears that we have no exact theory for skew-bevel teeth that are of practicable utility, except at or near the gorge circles, those selected at a considerable distance from the gorge being worthless from excessive flatness of teeth, and doing more crowding than driving. Good practical skew bevels must, therefore, come from the approximate solution of Fig. 195, or be fitted with the assumed and conjugated teeth of MacCord.

It is plain that the velocity-ratio changes with a change of the inclination of the lines on the plane *G*, Fig. 202. This controverts the statement in Prof. MacCord's *Kinematics*, page 365, respecting twisted skew wheels.

The Olivier Skew Bevels find application in such examples as Figs. 204 and 205. To design those of Fig. 204, draw the axes *AO*

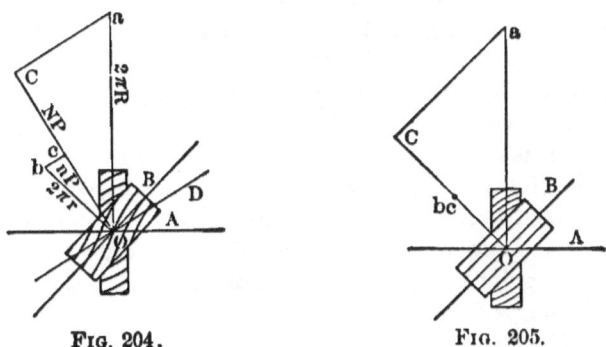

Fig. 204.   Fig. 205.

and *BO*, and generating line *DO*, as answering to one of the lines on the plane *G*, Fig. 202. Then on a line perpendicular to *DO* lay off the distances *OC* and *Oc*, equal to the number of teeth *N* and *n* in the large and small gears respectively, multiplied by the normal pitch. Draw the triangles *OCa* and *Ocb*, right angled at *C* and *c*, and with *Oa* and *Ob* perpendicular to the axes. Then the diameters or radii of the wheels will be in the relation of the lengths *Oa* and *Ob*.

In Fig. 205 one wheel is a spur gear, answering to the case in the Olivier gears where the lines on plane *G*, Fig. 202, are parallel to one of the axes. The diagram is lettered for similar quantities as in Fig. 204.

To cut these teeth, so as to make the wheels Olivier wheels, the cutters must be of such special forms as to give to the teeth, in sections normal to the axis, the involute form outside the pitch lines. But this is hardly to be expected in practice, and ordinary cutters will be used, even if the teeth have contacts at points only.

In Figs. 204 and 205 either or both wheels may be enlarged to the rack, which in action will slide endlong in their own tracks.

The Worm and Gear may be made as Olivier gears by making one wheel very much smaller than the other, and giving the smaller one only one ridge or screw thread, by properly inclining the lines on plane G, Fig. 202, and by use of the proper shaped cutter.

But this, again, can hardly be expected in practice, as good working worm wheels are produced by using a hob cutter of the same size and shape as the worm itself to finish cutting the gear.

The normal pitch should be the same here at the pitch lines for the worm as for the gear. When the axes are at right angles, as usually the case, the pitch of the worm in a direction parallel to the axis is the same as the circumferential pitch of wheel at pitch line. The wheel teeth, instead of being cut straight, are concave, as formed by the revolving hob cutter, thus securing more bearing surface between teeth.

In a section taken by a plane normal to the wheel axis and containing the axis of worm, the shapes of the teeth of worm and wheel should be the same as in the rack and pinion of Figs. 162 or 167, with the same points observed as to interference of teeth. The worm may be single or many threaded.

The Hindley or "Corset-shaped" worm may be supposed to possess advantages over the cylindric worm when examined in section; but a little consideration will show that its tooth contacts with the mating wheel are more like points, or, at best, a line on each tooth from top to bottom, because of the continually varying diameter of the worm, so that no thread can fit the teeth of the wheel as well as the cylindric worm thread can, with line contacts along the thread from side to side of wheel; and second, that the larger average diameter of the worm will give rise to more rapid wear.

Skew-bevel Pin Gearing is possible for all angles between axes, but the pins take complex shape, except for the one case of right-angled axes and velocity-ratio unity; when the pins are cylindrical and in diameter equal the shortest distance between axes.

Worm pin gearing is also possible; but as these pin wheels are scarcely demanded in practice, they will not be described here. Detailed descriptions are found in MacCord's *Kinematics*, pp. 284–293.

Intermittent Motions.

Movements of this class are given in Figs. 14 to 16, where the pitch lines were discussed. The teeth on these wheels are the same as those already considered, for both the principal circular arcs and the non-circular initial arcs, and also for the starting and stopping accessories, as explained in Figs. 140 to 143.

These movements may be made for the three cases of (1) **Axes Parallel,** (2) **Axes Meeting,** and (3) **Axes Crossing Without Meeting;** but, as the principles for all these have been duly considered for non-circular pitch lines, they will not be taken up here for the simpler case.

CHAPTER XVII.

ALTERNATE MOTIONS.

I. LIMITED ALTERNATE MOTIONS.

DIRECTIONAL RELATION CHANGING.

First. With Solid Engaging and Disengaging Parts.

In Fig. 206 is illustrated a movement of this class with engaging and disengaging features, all designed with a view of being a thoroughly practical and durable working movement. It is laid out with involute teeth.

If revolving right-handed, before the tooth c can strike upon e, the tooth a will have moved to some extent upon its mate b and to

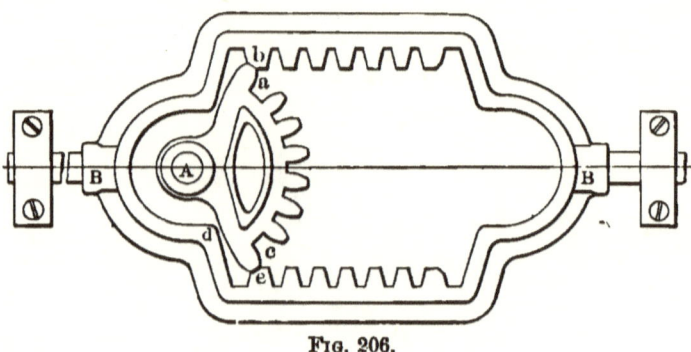

FIG. 206.

have thrown the rack fully, so that interference cannot occur at ce. Then, to start B gradually, the portion at d is carried up near to the center of A, so that the first impulse for moving B is received at d, where the velocity of A is less than at a, and thus reducing the initial blow. This may be carried still farther toward the axis A if desired.

In this construction there will be a momentary pause of the rack at the end of the movement.

Second. With Attached Engaging and Disengaging Spurs.

In Fig. 207 the movement has attached spurs to control the reversal of movement so that the blow may be reduced. This is taken from a working design. The teeth are epicycloidal, though they may be of any form preferred.

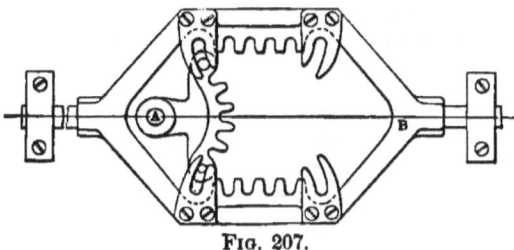

FIG. 207.

The shock due to initial motion is reduced to a minimum by carrying the spurs up so far that the normal to the curves of spur and pin will strike so close to the axis of A as to indicate easy starting.

For simplicity, the movement of Fig. 206 has the advantage, though this will be accompanied with less shock at reversal. The spurs may be carried still higher to farther reduce the shock.

The Mangle Wheel.

Under velocity-ratio constant the pitch lines must be circular, and in this movement the wheel is limited to one revolution at most, an example of which is shown in Fig. 104.

In this form of axes meeting the wheel may have ordinary teeth on the edges of a band instead of pins; but for a face wheel pins would be necessary, otherwise the opposite sides of a band bearing teeth would give different radii of pitch lines, and hence different velocities for forward and return motion. This construction may, however, be required, giving a case of velocity-ratio changing, though constant for each direction.

II. UNLIMITED ALTERNATE MOTIONS.

These are found in the mangle rack, which may be conceived of by supposing the wheel of Fig. 104 to be cut at the center of the U-shaped reversing piece, then straightened out to a rack, and then mounted on proper slides.

The teeth may be on the upper and lower sides of a rack bar instead of a row of pins. Any length may be given to the movement, and hence it may be classed as unlimited.

CHAPTER XVIII.

CAM MOVEMENTS.

IN a cam movement the path and the law of motion of the follower may be determined independently of the driver, when the latter is made to conform therewith.

Thus the follower path may be a straight line, a circle or any other curve, and the movement of the follower in that path may be assumed point by point for the full forward and also for the backward movement.

The cam is accompanied by an undue amount of friction by the rubbing of the follower on the driver, which in turn contributes to the wear of the parts, causing backlash and, in high speeds, noise.

This movement is the one usually called to his aid as the last resort, when the designer, seeking to avoid it, fails to obtain the necessary motion of a piece by other means, such as may be proposed for lighter and more quiet running of parts.

The cam is, therefore, a most useful movement, and in certain cases must be accepted, though to be avoided whenever possible.

CAMS IN GENERAL.

BY CO-ORDINATES.

The driver may have a non-uniform motion as well as the follower. To unite all questions in a general solution of a cam and follower in one plane, let A, Fig. 208, represent the driver axis, to which is attached a pointer, $A1$, the same turning so as to be in positions 2, 3, 4, etc., at the ends of equal successive intervals of time. The unequal angles at A indicate variable motion of A. Let the curve 1, 2, 3, etc., at D represent the path of the follower, D the point of the latter, which reacts against the cam, being at positions 2, 3, 4, etc., at the ends of equal successive intervals of time, these intervals being the

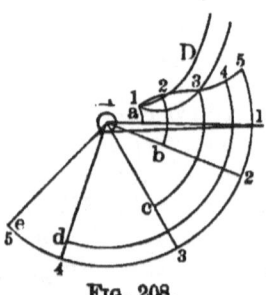

FIG. 208.

same as for A. The mechanism by which D is mounted, compelling it to move in the path 1, 2, 3, etc., is not shown.

Draw a circle arc $1a$ from the follower path to the position 1 of the driver pointer. Likewise arcs $2b$, $3c$, $4d$, etc., as shown. This gives us co-ordinates by which to construct the cam, as in Fig. 209, where $a1$, $b2$, $c3$, etc., are laid off from the initial line $A1$.

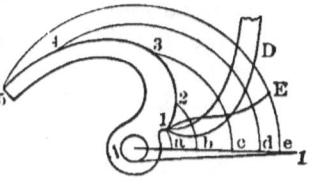

FIG. 209.

Then the cam outline may be drawn in as the curve 1, 2, 3, 4, etc. In practice this may all be done in the same figure.

This cam, $A5$, while turned from the position shown around till 5 comes to E, will evidently drive the follower's reacting point along in its path from 1 to E. The arc $E5$, from the construction, evidently equals the arc $1e$ of Fig. 208, as it should, and likewise for the other arcs. Hence the follower will move in its path as proposed, while A has the motion assigned to it.

By Intersections.

The cam may be drawn by the method of intersections, as in Fig. 210, where the follower path is laid off in the several positions in the inverse order of motion of A, following the successive equal intervals of time, as shown.

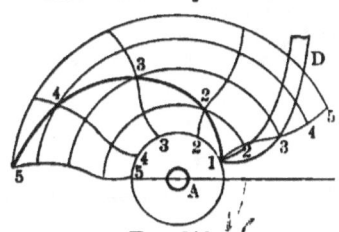

FIG. 210.

For convenience in this, a templet may be cut to the follower path curve, with center point at A noted. Then with the angles $1A2$, $2A3$, etc., laid off and noted, the several curves can be struck by templet.

Now, drawing in the circle arcs from the points 1, 2, 3, etc., of the follower path, we obtain intersections and can draw the curve 1, 2, 3, 4, etc.

The method of intersections is usually employed in practice, and as the motion of the driver A is generally uniform, the path positions 1, 2 2, 3 3, etc., are uniformly distributed.

Directional Relation and Velocity-Ratio.

In the above cases the directional relation is constant but for the follower to return to its starting point, as A continues, it must be changing. For this we may draw the cam in the same way, laying off another set of points in its path for the return of the follower, and the corresponding angles of A.

When the angles for A are equal, and the points 1, 2, 3, etc., equidistant, the velocity-ratio is evidently constant, regarding the follower as that which moves in the path 1, 2, 3, etc.; otherwise not. Sometimes the follower moves in a straight line, and sometimes swings about a center B, the velocity of which in either case is to be compared with A.

It seems unnecessary to classify cams under directional relation and velocity-ratio, and they will be treated here without it.

VELOCITY-RATIO FOR A SWINGING FOLLOWER.

In Fig. 211 we have a cam A, and a follower BD, which swings about an axis B. The path of the point D which acts against

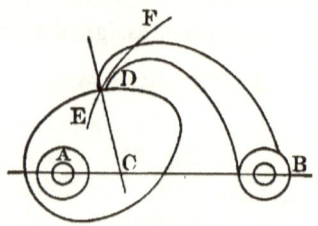

FIG. 211.

the cam is a circle arc, EF. To find the velocity-ratio between A and B, draw the normal DC, and then, according to Figs. 105 and 106, the

$$\text{velocity-ratio} = \frac{\text{ang. velocity of } B}{\text{ang. velocity of } A} = \frac{AC}{BC},$$

being thus in the inverse ratio of the segment of the line of centers, as has been found for other movements.

When the highest point of the cam passes under D, the normal CD will strike at A and the angular velocity of B will be zero. As A moves on, the point C changes to the other side of A, and the motion of B will be reversed, thus changing the directional relation. When C is between A and B the directions of motion are opposite, while for C outside the directions are alike.

VELOCITY-RATIO WHEN THE FOLLOWER MOVES IN A STRAIGHT LINE.

When the follower point D moves in a straight line, B in effect is at an infinite distance away, as in Fig. 212, and $BD = BC =$ infinity. Then the linear velocity at D is to be compared with the angular velocity of A for velocity-ratio.

To find the linear velocity of D, we have from the above the

linear velocity of $D = BC \times$ ang. velocity of B,

and this velocity-ratio, Fig. 212, will be

$$\frac{\text{linear velocity of } D}{\text{ang. velocity of } A} = AC.$$

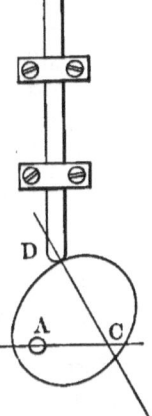

Fig. 212.

To draw an interpretation from this, take the expression in the form

linear velocity of $D = AC \times$ ang. velocity of A.

This signifies that the linear velocity of D in its straight path, and for the position considered, is equal to the velocity of the point C as if it were fixed upon the cam A and revolving with it.

CONTINUOUSLY REVOLVING CAM, AND RETURNING FOLLOWER.

BY METHOD OF INTERSECTION.

Assume the motion of A uniform, and that the follower swings about an axis B, Fig. 213, causing D to move in a circular path, D, 1, 2, 3, 4, forward, and 5, 6, 7, 8, D on the return. Draw circles through these points. There being nine divisions in the follower path, make also nine corresponding equal divisions in a circle struck through B. From each point of division as a center and with a radius BD strike the nine circle-arc follower-path lines 1, 2, 3, etc., as shown, intersecting the parallel circular lines drawn on the cam from A.

Then draw in the linear cam outline through the points of intersection of these lines of like number.

Numbers in the circles about A are in order left-handed. Therefore the cam is to revolve in the inverse order, right-handed.

Here the follower moves forward and then back to the point of beginning in one revolution of A, and the movements may be repeated indefinitely. The forward movement of D is faster than the return, there being four and five divisions to each respectively. These divisions of path are to represent the velocity of D.

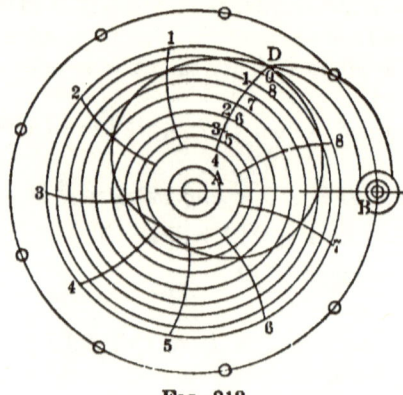

Fig. 213.

Here the directional relation is changing and the velocity-ratio varying.

Case of a Cylindrical Cam.

Here AA is the elevation of the cylindric cam and O the plan showing eight equal divisions, as in Fig. 214. At EF is the de-

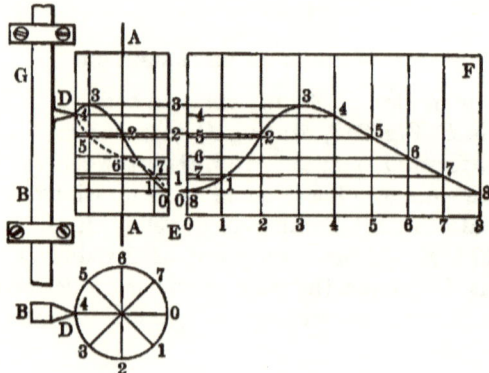

Fig. 214.

velopment of the cylinder, showing the linear cam, the same being also shown on the cylinder.

D is the point of the follower moving in a straight path parallel to the cylinder, so that the follower-path lines in the development are straight and vertical. Above E are laid off the points 1, 2, 3, 4, etc., in the follower path, and through them are drawn the parallel cam lines. Then the linear cam line is to be drawn through intersections of lines of like number, as shown.

By redeveloping upon the cylinder, we obtain the cylindric linear cam. In practice, the drawing may be made upon the cylinder itself direct.

The case of the figure is a simple one of a right-line movement of the follower.

The double screw movement is an example of a double cam motion, where a reciprocation is effected by several turns of the screw cam, the threads or cam grooves returning and crossing themselves. The follower in this case is oblong, and of length sufficient to guide it in one groove as it passes another.

In case of a circular path for D, the vertical or follower-path

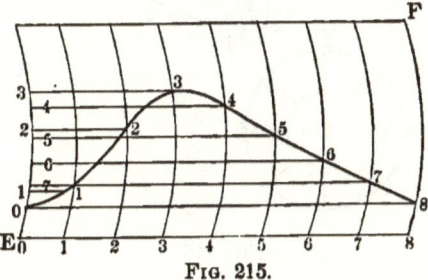

FIG. 215.

lines in the development would be circular and copied from the path, as shown in Fig. 215.

It is possible for the path of D to be some other curve, in which case the path lines on EF should be copied from it.

Case of a Conical Cam.

This is similar to Fig. 214, the parallel cam lines around the conic surface developing in circles as in Fig. 216. The follower path is here taken as an element of the cone for a simple case. But the path of D may be a circle arc or other line when the follower-path lines of EF will of course be drawn to the same circle arc or other curve as that of the path.

Case of a Spherical Cam.

Here the sphere AA is to serve for the cam, as well adapted for the case of axes A and B at right angles and intersecting, the

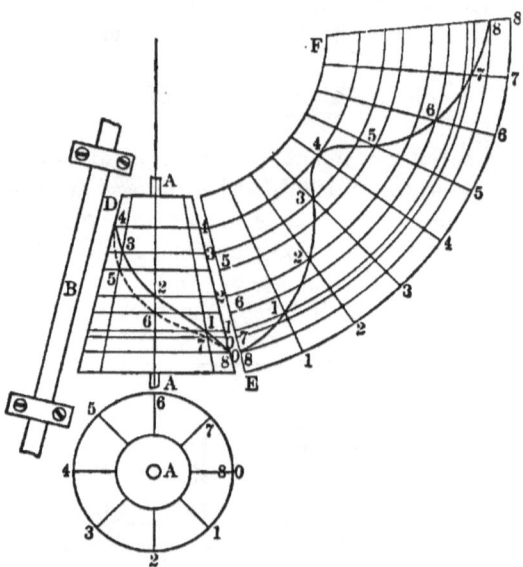

Fig. 216.

sphere being centered at the point of intersection, as in Fig. 217.

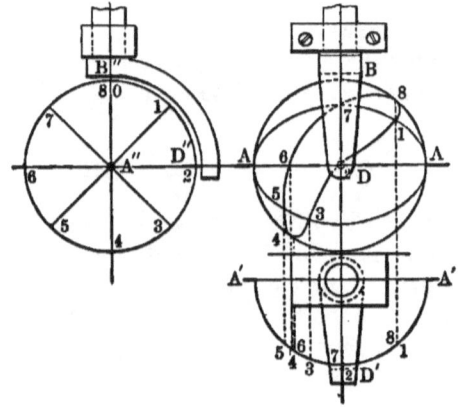

Fig. 217.

The position of the follower point D is so chosen that the angle

$D''A''B''$ is a right angle, and the motion of D in its path will describe a meridian to the sphere. Then the proper number of meridians and parallels are to be drawn for path lines and cam parallels, the latter through the points 1, 2, 3, etc., of the path, as seen in plan at $A'D'A'$. The linear cam may now be drawn as before, through intersections of lines of like numbers.

It is easy to imagine a less simple case, as, for instance, where the angle BDA is not a right angle for which a follower-path line of D on the sphere would not be on a meridian, but on an oblique great circle.

Also, the line of B might be at a considerable distance back or in front of the axis AA, as for the case of *axes crossing but not meeting*, when the cam blank ADA will be in the form of a circular spindle, or a circular zone, respectively.

A Plane Cam or Cam Plate.

For this the cam will be like a sliding plate, a cam plate; as if the development of Fig. 214 or 215 were mounted to slide on guides in reciprocation, or the corresponding part of Fig. 216 to swing about a pivot.

Cam with Flat-footed Follower.

In some cam movements the follower has a flat bearing piece, D, Fig. 218, instead of a point, which for the same cam changes the law of motion of the follower, but it presents a more extended bearing surface to the cam. It is evidently immaterial where the guide rod B is, whether to right or left, provided the pressure at D does not cramp B excessively. Again, D may be mounted to swing about a center, B', as shown in dotted lines.

The velocity-ratio for B is to be found as before, with respect to the normal DC.

Cams for Specific Law of Motion of Follower.

1st. A Uniform Motion of Follower, in Reciprocation in a Straight Line.

In Fig. 219, take D, 1, 2, 3, 4 as equidistant points in a straight follower path which passes the axis A at a distance EA from it. Draw parallel cam circles through the points 0 to 4, and the follower-path lines as tangent to the circle EA, cutting the outer circle at equidistant points, 1, 2, 3, 4, etc., to 8.

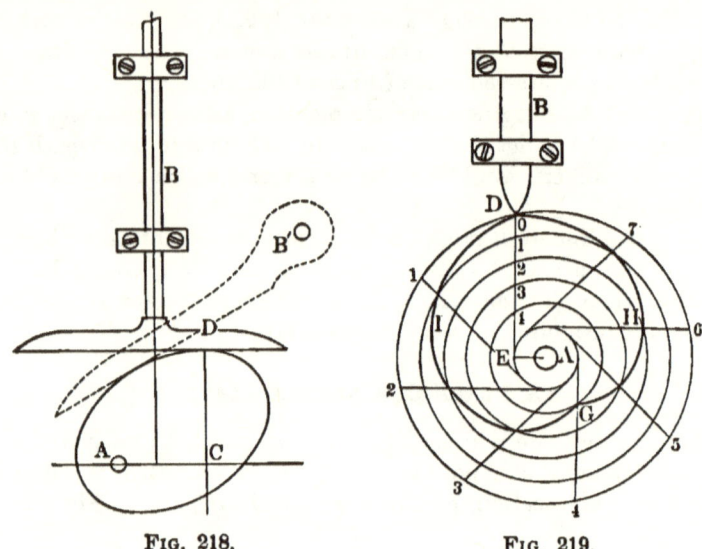

Fig. 218.   Fig. 219.

Then the linear cam can be traced in as shown, which, as a cam acting on D, will impart to it a uniform motion, 0 to 4 and 4 to 0, with the same constant velocity-ratio.

The curve DHG will be an involute to the circle EA when the length of path is equal the half circumference of EA, and is the correct cam curve for the lifting cams of a miner's stamp mill, the stamp head rods being in the line ED.

The velocity-ratio being constant, the normals to the linear cam curve DHG, at any contact of D, will all intersect the line EA produced in one and the same point E at the left of A. Also, the normals to the cam curve DIG will intersect the line EA at the same distance from A on the opposite side.

If the construction be such that the point E coincides with A, the linear cam curves become Archimedean spirals, and the cam one of constant diameter, and known as the heart cam.

2d. Law of Motion that of the Crank and Pitman.

In Fig. 220, A is the center of the cam, O the center of the crank shaft, and $abcde$ half orbit of the crank pin, in which the

CAM MOVEMENTS. 201

latter is supposed to move at a uniform rate, making the spaces *ab*, *bc*, etc., in equal times.

The pitman has the length *ao*, *bD*, etc., *e*4, so that the spaces 0 1, 1 2, etc., are passed in equal times by the end *D* of the pitman *bD*, and with a varying movement in accordance with the so-called crank and pitman motion. Through these points draw the par-

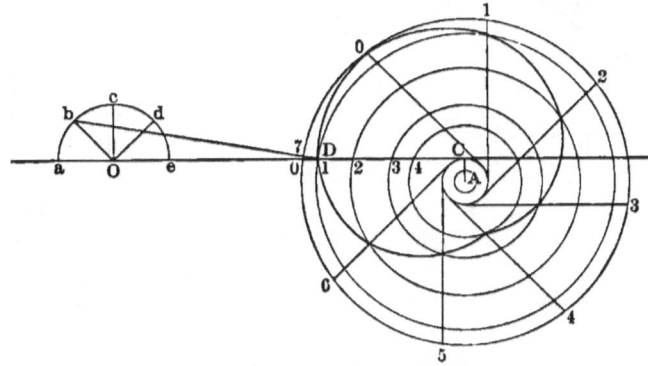

FIG. 220.

allel cam circles and divide them into equal parts by the follower-path lines as tangents to the circle *AC* and to equal points of division, 0, 1, 2, 3, etc.

Then draw in the linear cam, through points of intersection of path and cam circle lines of like numbers. This cam will give the follower *D* the same motion as would the crank and pitman; the crank or the cam both being supposed to revolve with uniform motion.

3d. Law of Motion that of a Falling Body.

In Fig. 221, *A* is the center of the cam, *D* the follower to move in the path *D*4, its velocity in the first half of which path is to be accelerated as that of a falling body, and in the last half to be retarded according to the same law.

To express this law on the line *D*4 of the path, draw a parallel *ag*, on which as an axis draw a parabola *abc*, and an equal one, *gdc*, intersecting it at *c*. Make several equal divisions *gf*, *fe*, etc., and from these point off equal divisions, erect ordinates *ec*, *fdb*, etc., and from the points of intersections, *d*, *c*, *b*, etc., draw horizontal lines over to the path *D*4. Then if the spaces *aj*, *ji*, etc., or 0 1,

1 2, etc., down to the middle of *ay* are described in equal times, the law of motion is that of a falling body, from the known property of the parabola. The reversed parabola *gdc* will give the desired points in retardation.

One peculiar property of the action here is that the inertia of the parts of the follower will cause a constant resistance in the follower path—a probable advantage where the follower is very heavy.

In practice, numerous lines should be used in the construction, a few only being employed here for clearness of figure.

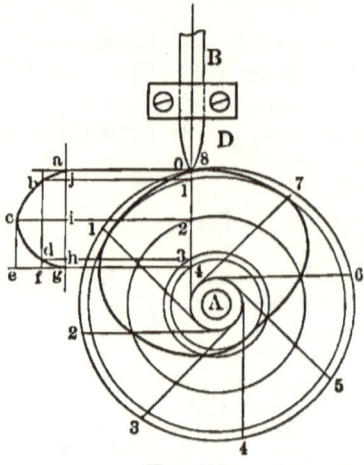

Fig. 221.

4th. Tarrying Points for Follower.

It often happens that the follower is to make a movement, then halt momentarily, then make another movement, then halt, etc. These halting or tarrying points are easily provided for in the cam by introducing circular segments, as shown in Fig. 223.

5th. Uniform Reciprocating Motion.

The so-called heart cam is a common device for this, an example of which is shown in Fig. 222. As a linear cam it has a constant diameter through the axis, and when the follower is located to coincide with this axis and is provided with a bearing point against the cam, both above and below it, there will be no backlash between the cam and follower.

Fig. 222.

6th. The Heddle Cam.

This can be used to work the heddles or harness in looms where the intervals of motion and rest are about equal. Fig. 223 illustrates, by **photo-process** copy, a working cam of this kind.

To make a heddle cam of constant diameter, uniform motion, and with tarrying intervals of time equaling the motion intervals:

In Fig. 224 divide the cam into four equal parts, retaining the opposite segments DH and EG for the tarrying intervals. The remaining opposite segments must serve for the linear cam lines. In these segments draw equidistant circle arcs, and also radial lines, in equal number. The linear cam lines may then be drawn in through points of intersection as shown, giving us the linear cam outline $DbEGeHD$.

Fig. 223.

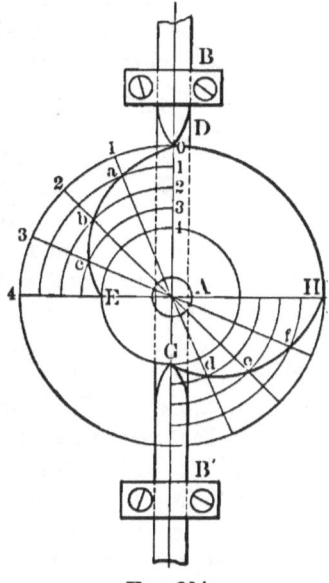

Fig. 224.

All lines of this cam through A will have the same length, or the cam will have a constant diameter, and it will just fill the space between follower points D, G on a slide BB'.

The curves DbE and GeH are Archimedean spirals.

7th. Easements on Cams.

The cam of Fig. 224 will instantaneously start the follower into full motion, the latter remaining uniform until stopped with equal abruptness. This is objectionable in any case, and when the follower is heavy, it is destructive. This is obviated by what may be termed easement curves for the cam, or easements.

An Arbitrary Easement may be drawn, as in Fig. 225, where A is the center of the cam, and G a point of abrupt change between the tarrying and cam arcs.

Assume equal angles EAG and GAF, and divide them into

equal parts by lines extended outward. Then note points *a*, *b*, *c*, etc., over to the line through *G*, and in increasing distances from *EG*, such as estimated to give a proper easement curve when drawn through the points. Lay off the distances *fa*, *gb*, etc., on *ie*, *hd*, etc., until the center radius *Gc* is reached. Then draw the easement curve *EabcdeF* through the several points.

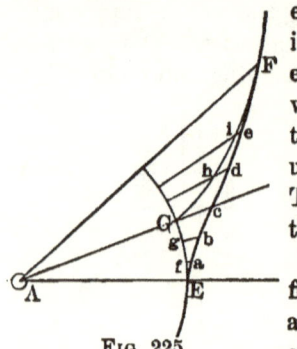

Fig. 225.

Treating the cam, Fig. 224, in this way, filling in at *E* and *G*, and taking off like amounts, on like radii and angles at *D* and *H*, we modify the cam, Fig. 224, by supplying easements, and without altering the constancy of its various diameters.

As thus treated, however, we encroach upon the tarrying arcs and shorten the times of rest of the follower. We may, however, provide for the angle *EG* in laying out the cam angles.

8th. **To Confine the Easements within the Assigned Cam Sectors**, take *DAH*, Fig. 226, as the allotted sector for the linear cam arc sought, including the easements. Let it be assumed that the easements shall start the follower as by the law of the crank and pitman motion; then, when full motion is attained, to continue that motion of follower uniform till approaching the arc of rest, and then to stop the follower by the same law of the crank.

To this end draw the quarter circles *DI* and *GF* from centers *K* and *J* on the follower path, and connect *G* and *I* by a common tangent. Then with dividers note points in equal division from *D* through *I* and *G* to *F*. Project these points to the path line *DF*, and draw circle arcs about *A*

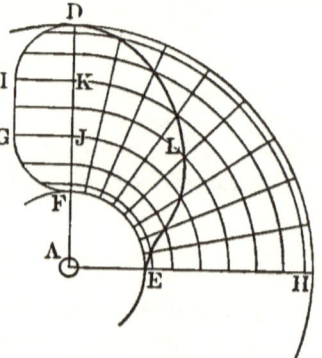

Fig. 226.

through the sector. Also draw an equal number of radial lines equidistant between *D* and *H*. The linear cam line *DLE* can now be drawn in, which will be found to have easements at *D* and *E* which are tangent to the arcs of rest for follower, and without encroachment thereon.

This linear cam will be found steeper at *L* than the cam of Fig.

224, and it may be a question which cam is preferable. But by making the circle arcs DI and GF smaller the easements may be more limited, and the inclination of the central portion of the cam arc less severe.

Other laws of easement curve may be adopted, as, for instance, DF may be taken as the base of a cycloid, in which the same number of divisions may be made as in $DIGF$. If that arc be shorter than the one here drawn, the resulting linear cam will have a milder declivity at L.

The use of any curve $DIGF$, which is symmetrical with respect to a perpendicular to the middle point of DF, will give a linear cam line DLE, which, if paired with a like one in the opposite quadrant, as in Fig. 224, will constitute a cam of constant diameter through A.

In Fig. 226 the law of motion of follower is represented by the spacing along the follower path DF.

CASE OF FLAT-FOOTED FOLLOWERS.

Let the law be that of a crank and pitman motion; determine a cam of the kind shown in Fig. 218. Draw the circle of the crank

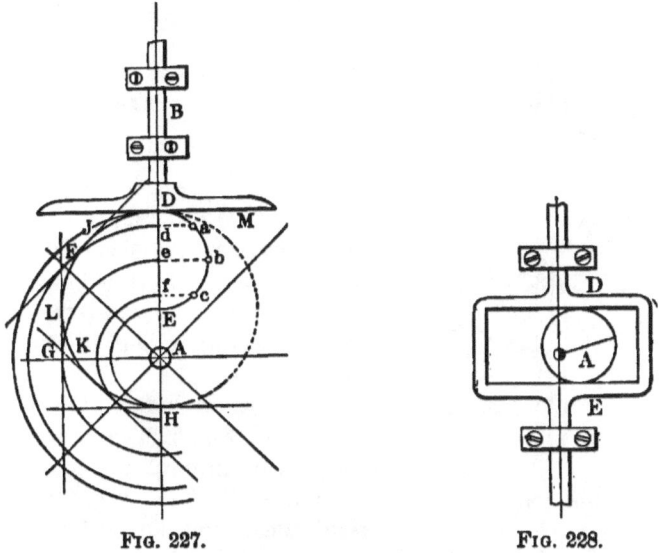

FIG. 227.   FIG. 228.

at DbE, Fig. 227, and project points a, b, c, etc., of division into equal crank arcs, to the radial line DE, giving points d, e, f, etc.

Draw the same number of diametric lines through A as FA, GA, etc., to represent the motion of A. Then draw cam parallel circles from points $d, e, f,$ etc. At intersections, as at F, G, etc., draw the follower line of DM as representing the position of DM relative to A when FA, GA, etc., have revolved to the line AD. That is, if DM is perpendicular to AD, draw FJ, GL, etc., perpendicular to the lines AF, AG, etc.

Tangent to all these position lines of MD draw in the linear cam curve, as $HLJD$. For this case the half may be copied on the other side of HAD as dotted, giving the same law of return of follower.

For this particular case and law, the cam is a circle or crank pin.

In cams with a follower consisting of a straight footpiece D, where the law on the return movement is the same as for the forward, the cam will be one of constant breadth as reckoned on a normal to the linear cam; and the follower may be made as a rack or frame DE, Fig. 228, with parallel sides.

Thus cams which while revolving just fill the space between a pair of points fixed upon the follower, as in Figs. 222 to 224, and 229, may be called cams of *constant diameter;* while those likewise just filling the space between parallel lines, or footpieces fixed upon the follower, as in Fig. 228, may be called cams of *constant breadth*.

CAMS OF CONSTANT DIAMETER AND BREADTH.

1st. Constant Diameter.

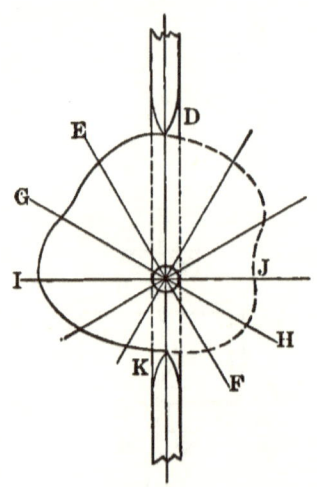

Fig. 229.

Besides those mentioned above, Figs. 222, 223, 224, 225, and 226, there may be cases where a specific law of motion is to be followed for a half revolution of cam, while the motion for the remaining half is immaterial as to law.

For this, as before, the follower points DK, Fig. 229, must move in a straight line through the center A.

Taking $DEGIK$, Fig. 229, as the essential cam for the half revolution, draw a series of lines, EF, GH, IJ, etc., through A, and make them all equal in length to DK, noting points FHJ, etc. Then the required counter half of the cam will be the dotted line $KFHJD$.

2d. **Cams of Constant Breadth** between parallel lines are possible, as in Fig. 230, where a half of A is assumed, and the forked

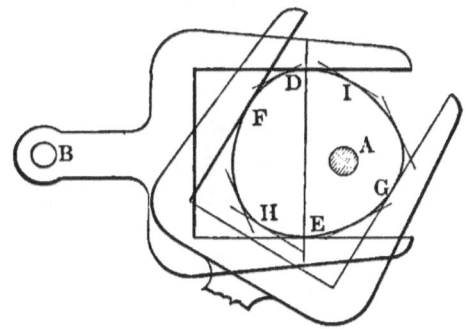

FIG. 230.

piece B to swing around the axis will just span the half cam at DE. Then by placing the fork in various positions as at F and G and drawing a line at G, again at H and I and drawing a line at I, etc., for a sufficient number of positions, and drawing an enveloping curve $EGID$, we have the remaining half of a cam, the breadth of which for all positions just fills the fork-shaped follower.

The outline of the cam, also its law of motion correspondingly, are limited, however, by the circumstance that the center of curvature of every part of the outline must be within that outline.

3d. **A Cam of Constant Breadth** as between parallel lines may be drawn by aid of a series of intersecting lines, as in Fig. 231.

Draw any system of intersecting straight lines. Then, beginning at some remote intersection as at a, draw a circle arc A from line to line, as those intersecting at a. This arc will be normal to the lines limiting it. Next, seeing that the arc B will meet the line intersecting at b, take b as a center and draw the arc B. The next arc will be C drawn from the center c, so that C will be normal to both lines intersecting at c. Draw the arc D from the center d; the arc E from e; the arc F from f; the arc G from g; the arc H from h; thus returning to the place of beginning. Arc H closes upon arc A exactly, if the work is right.

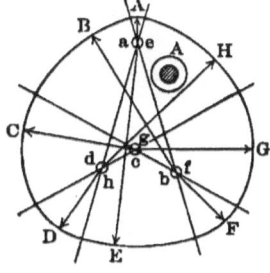

FIG. 231.

A little consideration will show that this cam, for every position, will just fill between a pair of parallel lines at a distance AE apart.

Also, it will just revolve, with a close fit, in a square hole of width AE each way; or in a rhomboidal hole of width AE each way, and with any angle between the parallel sides. Also, the hole may be bounded by parallel circle arcs, instead of parallel straight lines, partly or entirely, as in Fig. 237.

CAMS WITH SEVERAL FOLLOWERS.

1st. **One Cam may have Two or More Followers**, as shown in Fig. 232, the velocity-ratio of which may or may not be the same, as depending upon the construction of followers.

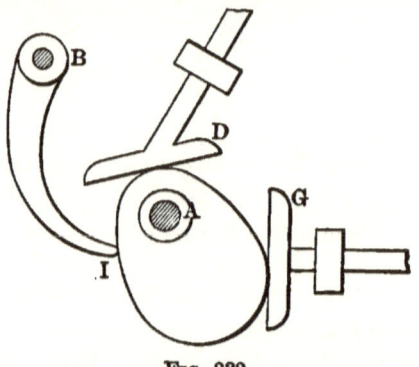

FIG. 232.

2d. **The Effect of Two Followers** may be obtained by combining two followers into one, as in Fig. 233, where the combined parts DG are solid, with a slide branch, E, the latter working in a box on

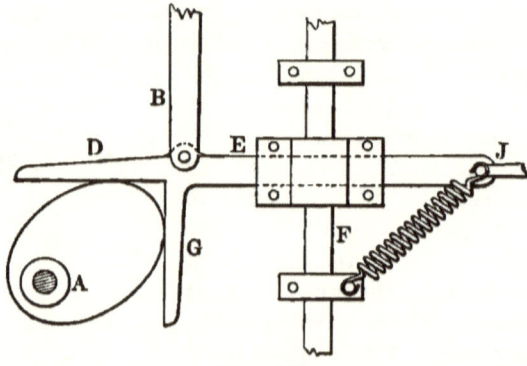

FIG. 233.

a second sliding piece, F, so that motion may be taken off at B and J in two directions.

3d. Fig. 234 gives a Similar Construction, in which DG swings about a pivot E, and that in turn about a second but fixed pivot F.

The follower connectors B and J may take the motion in two directions. One spring may serve to return the compound follower for the two directions of movement.

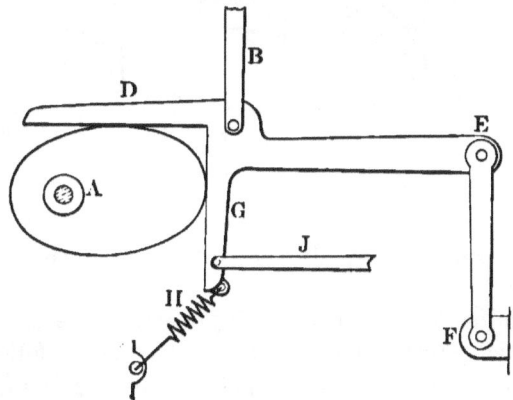

Fig. 234.

4th. Cams of Constant Breadth may return the compound follower positively, as in Fig. 235, where, as stated at Fig. 231, the cam of constant breadth will just fit and revolve in a square opening.

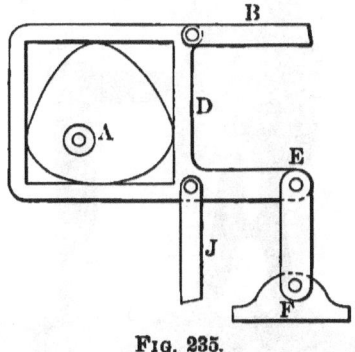

Fig. 235.

5th. The Four-motion Cam is shown in Fig. 236, and constructed by aid of intersecting lines in the manner shown in Fig. 231.

When the lines AF and AE are at right angles and EF intersects them at equal angles of 45 degrees, and the cam is drawn from centers AEF, as shown, the follower D, when pivoted at a relatively great distance to the right, will have every point moving in an exact square; as, for example, the point d will describe the

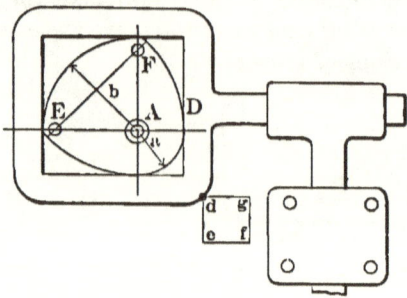

Fig. 236.

square $defg$, where de equals the difference of the construction radii a and b.

This cam has been used for sewing-machine feeds, for which it is well adapted, except that for this, one side of the follower should be cut out and an adjustable stop put to it to vary the feed.

In drawing the cam, it is immaterial as to motion how large it is, provided the periphery goes through or outside the points E and F, outside being preferable to avoid sharp intersections at those points.

In Fig. 237 is illustrated a working cam of this kind, where two

Fig. 237.

of the parallel lines of the follower are straight, and two are parallel circle arcs struck from the upper joint pin. This cam fits closely both ways in the opening, and the circle arc sides have the effect to keep the vertical connector bar quiet for two of the four movements, its motion being intermittent in reciprocation.

6th. **The Peculiar Shaped Cam, called a Duangle** by Reuleaux, is shown in Fig. 238. The duangle closely fits in the triangular opening in the follower for the entire revolution.

The follower moves forward and immediately back, and there tarries nearly stationary for about a sixth of a revolution, when another reciprocation is commenced.

FIG. 238.

An interesting double cam motion, which Willis would class as resulting in an aggregate path, is given in Fig. 239. Two arms work agreeably to the end of carrying a pencil point in such a manner that it writes the letters "O.S U." and places the periods. Model due to the ambitious energy of D. F. Graham.

FIG. 239.

RETURN OF THE FOLLOWER.

In most cases in practical machinery it is very essential that the return of the follower be positive in order to avoid breakages, though sometimes gravity or a spring may be employed where the velocity is not unduly high, and damage not likely to occur.

That the Action of Gravity may Quicken the Return of the end D, put the weight W near B, Fig. 240, and make the arm DW comparatively light, so that as W falls by gravity the end D will move more rapidly than W by reason of the leverage.

In this case take g, the "center of percussion" of the piece BD.

When *g* is in a horizontal line through *B*, *g* starts to fall just as

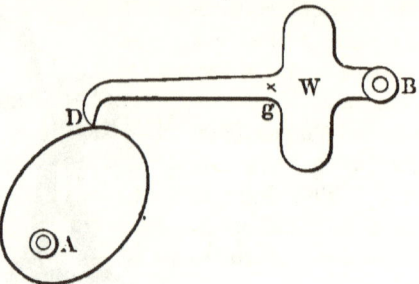

Fig. 240.

fast as a free weight, and the end *D* faster in proportion to its greater distance from the pivot *B*.

A Positive Return is readily obtained by making a groove for the end

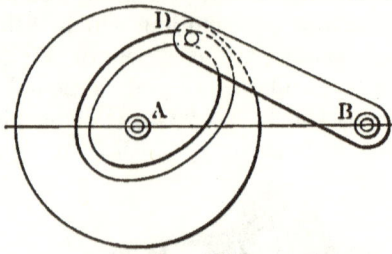

Fig. 241.

D to run in, so that it is compelled to move both ways, as in Fig. 241.

A positive return may also be insured by means of an extra cam,

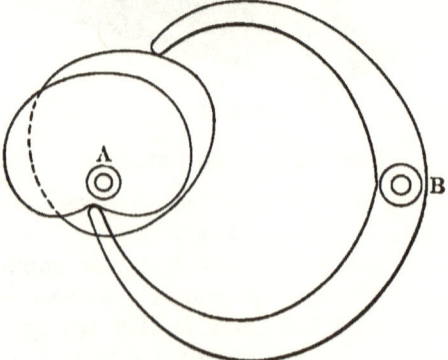

Fig. 242.

as shown in Fig. 242, one cam working in unison with the other to prevent backlash between either follower and its cam.

CAM MOVEMENTS.

In certain cases the cam may be made of constant diameter with the follower moving in a straight line, as shown in Fig. 243, when the follower points *DD* may just contain the diameter without backlash whereby the follower is compelled to move both ways. This may serve where a certain law of motion of follower is to be insisted upon for only a half revolution of *A*; or the same law for the forward and return motion of follower as in Figs. 222 to 226.

In cams of constant breadth the return of the follower may be assured, as in Figs. 228, 230, 236, and 237.

To Relieve Friction.

Where the follower is compelled to drag over the full extent of the periphery of the cam at each revolution extensive wear is very likely to occur. Lubrication will help, but it is difficult to maintain this on surfaces so exposed to air and to the tendency of centrifugal force to throw the oil off. The usual remedy is to put a roller at *D*, as in Fig. 244, which can track along on the exposed edge of cam, regardless of lubrication.

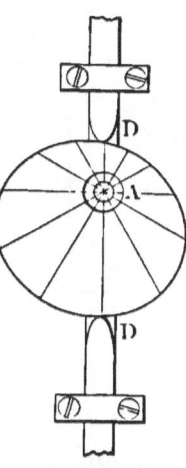

Fig. 243.

While rubbing of surfaces occurs at the pin there will be surface bearing, instead of a line bearing of a fixed terminal *D* on the

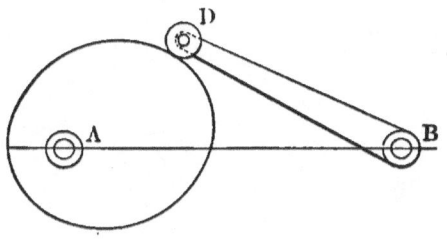

Fig. 244.

edge of *A*. Also there will be more favorable conditions of lubrication between roller and pin than between terminal and cam. The amount of rubbing action of surfaces will also be reduced in the ratio of the diameter of roller to diameter of pin.

In conical cams the roller is sometimes made conical.

In stamp mills a disk roller, Fig. 245, is used, which, though largely reducing the friction, does not do so as completely as the cylindric roller. Here A is the cam lifting the rod EF by acting against the collar D on the rod. The rod and collar both turn together as A may dictate. But owing to the thickness of A, with D rotating, the radius of D at one side of A differs from that of the other side, so that there is a combined rolling and torsional twist between the surfaces, and not an entire relief from slipping.

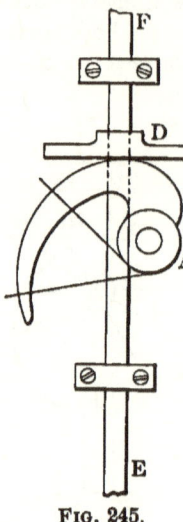

Fig. 245.

A flat-foot follower, as in Figs. 227 or 230, will be found better for endurance in wear than a sharp edge, as in Fig. 243.

Modification of Cam Required by Follower.

The immediate result of solution of a cam motion is usually a so-called linear cam, from which the practical cam is to be obtained.

In Fig. 246 the linear or theoretical cam is the curve $Dabc$, etc., which is the line as traced on A, which the theoretical follower point D is to follow as A revolves.

To Introduce a Follower Roller at D and maintain the theoretical action, the center point of the roller must follow the linear cam

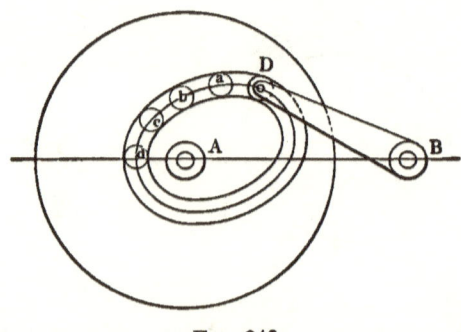

Fig. 246.

$Dabc.$ To insure this, strike circles equalling the roller in diameter at numerous points along the linear cam, as at a, b, c, etc. Then draw the practical cam lines tangent to these circles throughout.

CAM MOVEMENTS. 215

If these enveloping cam curves are drawn at both sides of these circles, we have a drawing of the cam groove for cam *A*, providing for a *positive return* of the cam follower.

In practice, this groove is best executed in metal, by using a cutter of the same shape and diameter as the roller *D*, and so mounting it in a cutting machine that the cutter while cutting is compelled to move with its center following the line *abc*, etc.

In Cams with Salient Angles, as in Fig. 219 and the heart cam of Fig. 222, a sacrifice must be made from the theoretical action of the cam by the introduction of a roller, as shown in Fig. 247. Let *FNEMG* represent the linear or theoretical cam and the line the center of the roller should follow. Drawing a series of roller circles, and the practical cam *HIJ* tangent to the circles, it is found that at *I* the cam is foreshortened by the amount *LE*. That is, for the practical cam *HI* to compel the roller center to move on the linear cam *FN* to *E*, the cam surface *III* would require to be extended to some point near *M*, where its normal would strike *E*. But as the cam is cut away beyond *I* by *IJ*, the roller is at liberty to swing around its point of contact with *I*, the center of the roller describing the circle *NLM* and failing to reach the point *E* in the linear cam by the amount *LE*.

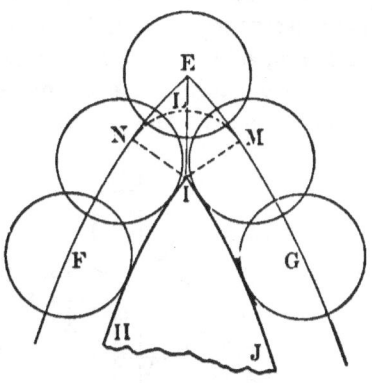

FIG. 247.

This circumstance may at times prohibit the use of the antifriction roller, as depending upon how essential the shortage *EL* may be. Or, at times, the roller may be used of diameter smaller than otherwise preferred.

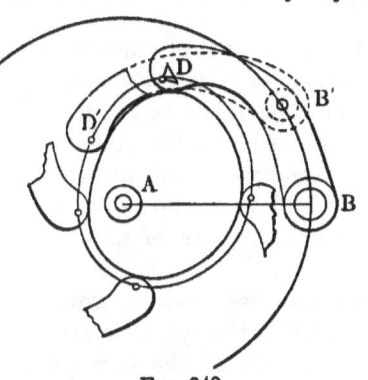

FIG. 248.

For a Thickened and Rounded Follower Extremity, *D*, as in Fig. 248, where some point *D* is to follow the linear cam as shown, cut a templet BD of the right shape at D for the follower, and with the point D notched by a V notch as shown, through which the linear cam can be seen. Then, with a circle about A through B, place the templet at numerous positions, as at $B'D'$, and draw a curve at the end, as shown. A cam line, traced tangent to all these curves as shown, will be the practical cam required.

For an Edge Cam, as in Figs. 242 and 244, and for a disk or face cam, with a cam groove in the side, as in Fig. 246, the roller should be cylindric without question.

The Best Form of Roller has been a matter of some question, as some designers use a cylindric and others a conic one for a cylindric cam, such as shown in Figs. 214 and 215.

With regard to the action on the outside surface of the roller alone, it appears that when the roller is moving longitudinally in

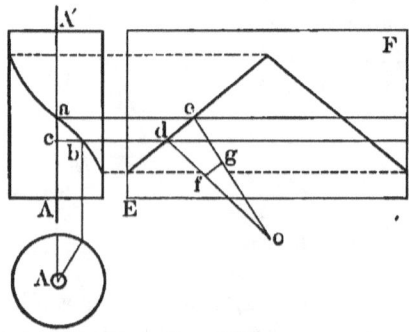

FIG. 249.

its groove, if such might be, its form should be the cylinder; while when the cam is revolving and the roller nearly stationary its form should be conic, with the vertex of the cone at the axis of the cam. For forms of the cam between the above limits the roller would seem to require some compromise form as between the above cylinder and cone. A little consideration will show, however, that a perfect form for simple rolling contact in this case, of axes of cam and roller meeting, does not exist.

Suppose the cam approaches one of the same velocity-ratio forward and back, as in Fig. 249, where AA' is the cylindric cam and EF its development. Take ab for one side of the cam groove, and de the same in the development.

Now, if the roller is made conical with its vertex at the axis A, when the roller rolls from a to b at the surface of the cam it would be necessary for the vertex to roll from a to c on the axis A, which is clearly impossible since the vertex is a point. From this it would seem that the roller must have some size at the axis; and probably the best that can be done by approximating it is to lay off the distance ac at fg, where fd equals the radius of the cylinder A; draw df and eg produced to meet in O, and take dO as the length of the cone from the large end of which to cut the follower roller.

Perfect rolling of the follower roller rolling along a cam surface ab would require that the axes of cam and roller pass each other with a distance between, and that the roller be a hyperboloid of revolution; also, that the distance between the axes varies as the inclination ab of the cam curve varies. It therefore seems impossible to obtain a perfect follower roller for a cylindric cam, that is, one where the outside surface of the roller simply rolls on the surface of the cam groove, because the action will of necessity be partly rolling and partly torsional slip of surfaces, and vary with slope of ab.

This torsional slip of surfaces will be the same for a truncated conic roller with vertex at axis of cam and moving in a cam groove parallel to the axis, as for a cylindric roller of equal length moving in a groove encircling the cam; and according to Fig. 249 these torsional slips for both rollers will be alike for a groove at an angle of 30 degrees with the circles of the cam surface.

The Action of the Roller upon the Pin and its Shoulders is of importance, as it is easily seen that the conic roller will be severely pressed against its shoulder, causing friction on a surface of larger diameter than the pin, thus introducing a very serious resistance to rotation of roller—an objection which, probably, by far outweighs the advantage of a conic roller, except in cases where the cam is of large diameter relative to the throw.

With regard to the pin for supporting the roller: when it can be made conical, with the same angle of convergence as the conic roller itself, the end thrust causing shoulder friction will be mostly avoided, and the conic roller will be much more acceptable. One drawback here, however, may be found in the large average size of pin, and it may be advisable to make it part way cylindric and the remainder conic.

For a Conic Cam the question of best form of follower roller is still more involved, and it is probably advisable to adopt the cylindric one.

For the Spherical Cam there seems to be no question but that the roller should be conical, since here the vertex of the cone can remain at the exact center of the sphere, and there will be theoretically perfect rolling between the roller and its cam groove. But even here the shoulder friction due to the endlong thrust will be found a serious objection to the conic form of roller unless the axial pin for the roller can also be conical to match, or partly conic and partly cylindric.

CHAPTER XIX.

INVERSE CAMS AND COUPLINGS.

I. The Inverse Cam.

THIS term may be given to a movement which has the elements of a grooved cam and follower, but where the driver has the pin or roller and where the follower has the groove; styled by Willis the *pin and slit*.

The inversion of the movement is to avoid dead points that would in some cases occur when used as a cam movement.

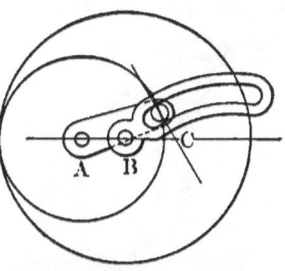

Fig. 250.

The peculiarity which distinguishes this from the cam movement consists in the fact that here the pin reciprocates in the slot or groove, while in the cam it does not.

The slotted piece B, Fig. 250, is the follower and A the driver with directional relation constant. The velocity-ratio as in cams is

$$\frac{\text{ang. velocity of } A}{\text{ang. velocity of } B} = \frac{BC}{AC}.$$

The slot may be made straight on a radial line or not, and a block may be fitted on the pin and in the slot to avoid wear by extending the bearing surface. Thus equipped with the slot radial, this movement is sometimes known as the Whitworth Quick Return. It is considerably used on English shaping machines.

An example of this movement for *directional relation changing* is given in Fig. 251. In this there is a point of tarrying of the driven piece by reason of the slot having quite a portion made to the circular path of the driver. Thus by curving the slot, modifications of motion may be obtained.

This movement has been used to give motion to the needle-bar of a sewing-machine.

220   PRINCIPLES OF MECHANISM.

Another example is given in Fig. 252 of a needle-bar cam motion much used in sewing-machines. It corresponds somewhat with Fig. 251, except that the axis B is in effect removed to infinity by the mounting of EF on a sliding bar.

At F the slot may be made to correspond with the circle arc described by the pin; so that the needle will

FIG. 251.

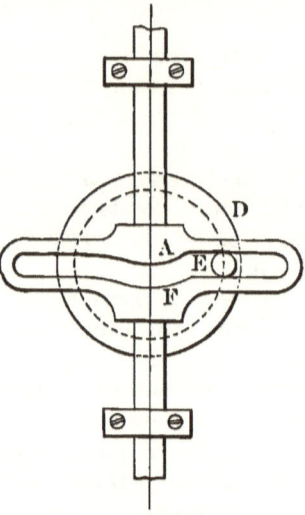

FIG. 252.

stand stationary while the shuttle passes the loop of thread. To form the initial loop, a quirk may be introduced in the slot at E.

In some applications in heavy machinery, this slotted piece EF has a straight slot and a block on the pin fitted to slide in the slot. Thus the pitman has been avoided in steam-engines and the engine correspondingly shortened.

By a sufficiently wide slot in FE, the movement may be placed at an intermediate point on a straight shaft and eccentric.

II. COUPLINGS BY SLIDING CONTACT.

Oldham's Movement with directional relation and velocity-ratio constant is illustrated in elementary form in Fig. 253 for connecting axes that are parallel but not coincident, acting by sliding contact.

The velocity-ratio is constantly equal to 1, as easily seen in the small diagram of a section normal to the shafts. The arms of the connecting cross which slide in the sockets made fast upon the shafts

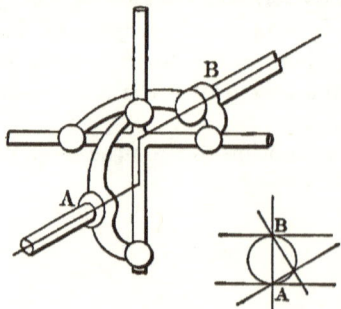

FIG. 253.

must constantly pass through both of the axial points A and B; and as they form a right angle, the point of intersection of the cross must follow the circle of diameter AB as shown, since all lines at right angles drawn from the extremities of a diameter meet in the circle to that diameter. The extent of sliding per revolution on each branch of the cross equals two diameters, AB.

If the cross is not right-angled, the same is true of the angular velocity, as shown in Fig. 254; but the amount of sliding is greater, since the circle is thrown to one side and increased in diameter.

A Complication of Movement results from placing the axes out of parallel, as in Fig 255, so that they meet at some point, O. Then ED shows one position of the cross, the angle being at E and F. Another position, a quarter-turn away, is

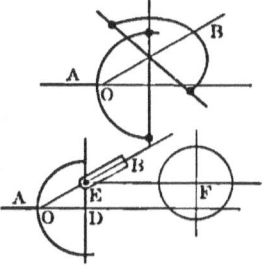

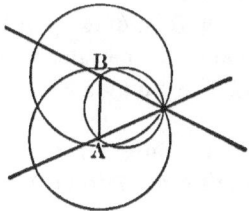

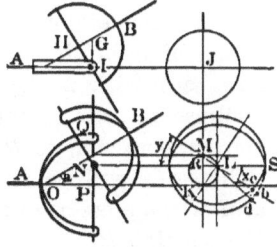

Fig. 254.   Fig. 255.

shown at III, with the angle at I and J. These diagrams, compared, show that the shaft B has endlong motion to the extent GH, twice in a revolution, regarding A as without end play. The angle point of the cross, however, always remains on the line IG, or plane normal to A, and describes a circle KLM in that plane.

To study the relative motions of the shafts A and B, take the cross as right-angled and in an intermediate position, KL, LM. The A-branch will always be found in the plane PN, normal to A, and the B-branch in a plane NQ, normal to B. These planes will intersect in a line perpendicular to the plane of the axes A and B, or in some line N, RL, for the position of the cross as shown. Draw a circle about L on the plane normal to A, and an ellipse to the same center, to represent a circle on the plane normal to B. Taking KL for the A-branch of the cross, and, as in the plane of the paper, the B-branch in its own plane will appear at ML with

MLK a right angle because lines at right angles in space will appear at right angles in projection, when one of the lines is parallel to the paper. Hence, if we move B from parallel to A into its

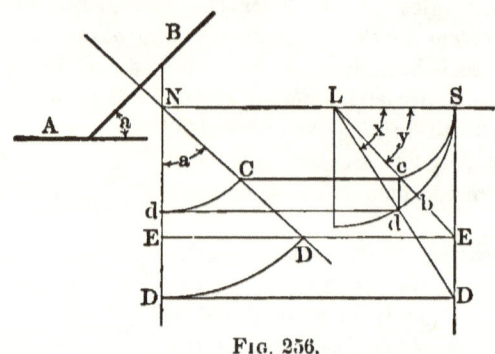

Fig. 256.

inclined position, while A and KL remain fixed, the point b in the circle must move to c in the ellipse. To determine the angular disturbance of B in this movement, swing B and its point, c in the ellipse, back, without angular disturbance, to parallel with A, when the ellipse returns to the circle and the point c will fall at d, cd being perpendicular to RL.

Then cLd will be the change in the angular position of the B-arm of the cross as due to the swinging of B from parallel to A, to the position BO.

Let x and y represent the angles SLd and SLb respectively and a the angle between the axes A and B. Then Fig. 256 will show the relation of these angles, from which we get $EN = DN \cos a$, $EN = LS \tan y$, $DN = LS \tan x$, which gives us the relation

$$\cos a = \frac{\tan y}{\tan x},$$

in which, a being constant, y may be found when x is given, or x found when y for any point in the revolution is given.

It may be noted that the same figure in cross-section at $RKLM$ is obtained whether the intersection is at O or at P; also the same equation. But it is clear that when O is at P the movement reduces to that of the Hooke's universal joint, which is free from the sliding motion of Fig. 255, and hence the latter can claim no advantages over the Hooke's joint.

The velocity-ratio could be found by calculating a series of angular positions from the formula and comparing them. But this is best done by differentiating the formula and obtaining

$$\text{velocity-ratio} = \frac{\text{ang. veloc. } A}{\text{ang. veloc. } B} = \frac{dx}{dy} = \frac{\cos^2 x}{\cos^2 y} \sec a$$

$$= \frac{1 - \sin^2 x \sin^2 a}{\cos a}$$

$$= \frac{\cos a}{1 - \cos^2 y \sin^2 a}.$$

When a equals 0 the angular velocities are equal, as they evidently should be, and the velocity-ratio $= 1$.

These equations are the same as given by Willis, p. 452, for Hooke's joint.

For $a = 45$ degrees; $\frac{dx}{dy} = \sqrt{2}$, or $= \frac{\sqrt{2}}{2}$, for maximum and minimum values as occurring for $x = 0$ and 90 degrees respectively.

The velocity-ratio is $\frac{dx}{dy} = 1$ for $\tan^2 y = \cos a = \cot^2 x$. These equations are the same as found by Willis and Poncelet for the Hooke's joint.

A Peculiar Movement transmitting motion from A to B is illustrated in Fig. 257, in which the contact between the parts is by sliding, except where the axes are in direct line. In the model, the axes may be arranged parallel or at various angles and at var-

Fig. 257.

ious offsets, as in Figs. 253 and 255, and it appears to be one way of realizing those cases in material form, except that here the axis B is not compelled to slide endwise.

To study the velocity-ratio it is most convenient to imagine the intermediate slotted piece to be replaced by a cross the branches of which are perpendicular to the slots they stand for. The movement will then serve for any and all the conditions brought out in Fig. 255 or 253 and with the same law of velocity-ratio.

These joints all have two points of maximum and two points of minimum velocity in each revolution, as also a pair of the two-lobed elliptic wheels of Fig. 44, but the law of velocity-ratio of the latter is different.

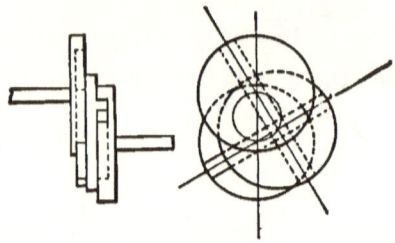

FIG. 258.

When the movement of Fig. 257 is set with axes parallel, it acts the same as the three disks in Fig. 258, which latter movement was used by Oldham in appliances employed in the Banks of England, by Winan in his Cigar Boat, and also by C. T. Porter to couple the shafts of an engine and dynamo nominally in line, but practically a little "off line" by reason of temperature, flexure, etc.

CHAPTER XX.

ESCAPEMENTS.

DIRECTIONAL RELATION CHANGING. VELOCITY-RATIO CONSTANT OR VARYING.

An escapement is a movement in which the follower is driven a distance, usually by sliding contact, to where the driver is allowed to pass free for a little space, when another engagement by sliding contact occurs to drive the follower back to a position for repeating an engagement like the first.

Power Escapements.

Fig. 259 illustrates an escapement of a design suitable for use in heavy machinery in which the motion is continuous and where there

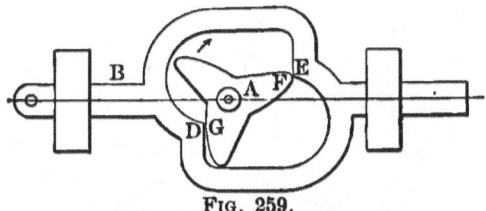

Fig. 259.

is no lock such as used in most cases of escapements for clocks and watches.

As F escapes from E, the arm G should be near to D to promptly engage; and the point of initial contact at D should be as near to the longitudinal line through A as possible, to relieve the blow due to initial contact.

Escapements are mostly employed in timepieces, and should be as nearly as possible such as to give to the vibrating pendulum or balance equal impulses at all times, and be as free as possible from hindering the vibration by frictional contact of parts with the pendulum or balance. That is, the higher essentials for fine timekeeping are, 1st, an isodynamic or equal-impulse escapement; 2d, an isochronous pendulum or balance; and 3d, freedom of vibrating

226　　　　　　PRINCIPLES OF MECHANISM.

parts from contact with the other pieces, except when receiving the impulses. There are other considerations, such as temperature, position, etc., which are outside of our present topic.

THE ANCHOR ESCAPEMENT.

In Fig. 260 is shown a so-called anchor escapement, the name being due to the resemblance of the vibrating piece to the ship's anchor.

As shown in full lines, we have the *dead-beat* escapement, in which the escape-wheel A stands still while a tooth rests on a pallet, notwithstanding the movement of the pallet. As shown, the tooth I is about to move forward upon the pallet HG, as the latter ad-

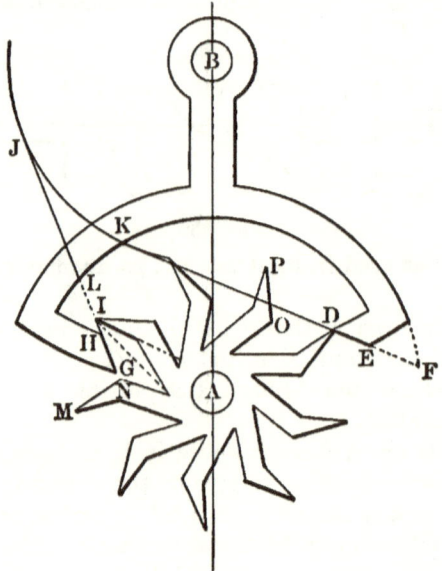

FIG. 260.

vances left-handed. Reaching H, a further movement of H to the right and return allows I to remain stationary, because the pallet from H back is formed to a circle arc about the center B. A like action occurs when the tooth rests and slides upon the pallet E. The teeth of A should be so formed and cut away as to permit the pallet to move a considerable distance after the escapement of a tooth and before the return of the pallet occurs, so that the pendulum may complete its swing.

Thus the tooth is just on the point of escaping at D, following which I moves forward upon the pallet H as the latter swings to the right, allowance being made for HG to move still farther toward A. As HG returns, the pressure of the tooth I upon the slope HG imparts an impulse to GHB toward the left. Similarly at D, the tooth is just completing its impulse on ED toward the right. These impulses overcome the retarding influences of the air and other resistances acting upon the pendulum, thus maintaining its motion.

To construct this movement, the pallets are somewhat thinner than the half of the pitch of the escape-wheel, so as to give a slight drop III to insure the landing of I upon II at a slight distance from the bevel HG. From II and G back, the pallets are formed of circle arcs struck from the center B. The bevel HG may be assumed by a line drawn to J tangent to a circle struck from B. The pallet ED should be formed to the same circles as HG, and beveled by a tangent to the same circle JK. Then the angle DBE will equal the angle HBG, as it evidently should to balance the impulses.

It has been proposed to put the impulse bevel upon the teeth of the escape-wheel instead of the pallets. Fig. 260 would nearly fit the case by turning A backwards when, as the tooth of D escapes, the tooth MN would fall upon the circle of the pallet G as an arc of repose, until, when the pallet returns, it would receive an impulse in sliding along the bevel NM until, when M escapes at G, O of the next tooth OP will land upon the pallet D, and in due time impart an impulse from the bevel OP.

Again, the impulse bevel may be divided between the pallet and the teeth of the escape-wheel, as really done in the ordinary " lever escapement " of watches.

The recoil escapement is obtained from the above by producing the bevel lines HG and DE, as shown in dotted lines, and modifying the teeth of the wheel to some shape as dotted at I. Then, as the tooth at D escapes from DEF, the tooth I will strike upon the bevel GHL, and, as the pendulum moves still farther in the same direction before returning, I will be forced to slide up towards L, thus giving to the escape-wheel a slight backward movement, called the recoil.

The recoil escapement is the most common one in ordinary mantel clocks, and regarded as inferior to the dead-beat, which latter is usually introduced into the finer mantel clocks, regulators, astronomical and many tower clocks.

THE PIN-WHEEL ESCAPEMENT.

This is so named because the escape-wheel has pin teeth, and it is shown in Fig. 261.

The pallets are formed to circle arcs ej, fi, gl, and hk, terminated with bevels ef and gh upon which the pins slide to impart the impulses to the pallets. A pin of the escape-wheel is shown as resting upon the pallet hk. If the pallets are moving toward the left, the pin slides upon the circle arc, or "arc of repose," toward k until the

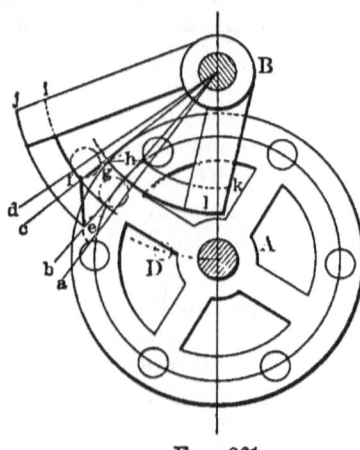

FIG. 261.

pendulum reaches its limit of swing, when it returns, and also the pallet, which, on arriving at the pin, permits the latter to slide down the bevel hg, imparting the impulse. It escapes from the pallet ghk and alights upon fi, of the pallet efi. The pendulum completes its movement and returns, when the pin slides down fe and imparts a second impulse opposite to the first. As the pin escapes at e, the next pin drops upon hk, to repeat the movements of the former one, etc.

To construct the movement, draw circles from the center B for the pallets, the latter having thicknesses such that the two, with a pin between, will swing between two adjacent pins of the escape-wheel, with a necessary slight clearance. Then the bevels ef and gh are so determined as to subtend the same arbitrary angle at B as bBc, besides allowing a small safety angle ab equal to cd, as providing for the distance from the impulse bevel back to the landing point of the pin upon the pallets. This angle may be small and possibly zero for cylindrical pins. When the pin escapes at e

(see dotted line), the next pin should land upon hk at the allowed angle aBb from h. Also, when the pin escapes from hg (see dotted line struck from D with a radius equal to distance from A to inside of pin) it should drop upon fi at the same angle cBd equal to aBb, from the initial point f of the bevel.

This escapement has been considerably employed for tower clocks, and has the advantage that the pins may conveniently be hardened, or even made of glass rod or cut jewel stones. In some cases the upper half of the pins are cut away, since this portion is not acted upon by the pallets, thus permitting the placing of adjacent pins a whole pin diameter closer. Then with the same number of pins and thickness of pallets, the wheel A will be made smaller, and the strength of an impulse will be materially increased other things the same. Also, the drop of a pin on escaping will be reduced by nearly a half, and the "tick" will be materially quieted.

THE GRAVITY ESCAPEMENT.

This name is given to escapements where the fall of a weight through a definite height imparts the impulse, all impulses being thus equalized in intensity. A spring may be used instead of gravity to measure the definite impulse. Such are sometimes called isodynamic escapements.

A gravity escapement employed by Wm. Bond & Sons in chronographs, and called isodynamic, is illustrated in Fig. 262. The same has been used in tower clocks with good results, and it is believed to be about the best in kind and in construction for that purpose, being the same in principle as the Bloxam's or Dennison's.

At A is the driving axis, on which is made fast a collar with an eccentric pin, a, and an arm, GH. A T-shaped gravity piece, DEF, is suspended by a spring, N, from a fixed clamp, QR, and will swing right and left, and it may rest against the eccentric pin a, or it may rest by its pin at F against the pendulum rod P. At D is a pin flattened on the lower side normal to a line DN. A second T-shaped gravity piece, IMK, is suspended on the other side and like the first except they are rights and lefts, and the pin at I is flattened on the upper side normal to a line IS.

The pendulum rod P is suspended from the clamp QR by a spring, not shown, which allows it to swing to the right and left.

The eccentric pin a throws the gravity pieces to the right or left. As shown, this pin a holds the gravity piece DEF in the extreme position the pin can give it; and the shaft *A* is locked in that position by the end of the slender arm *GH* striking against the detent pin at *D*. Now, as the pendulum rod, moving toward the left, strikes the pin at *F*, the gravity piece *DEF* is carried along with the pendulum to its limit of movement, thus releasing the arm *GH*, when, on making a half-turn, it is arrested by the end of the arm *GH* meeting the detent pin *I*, which now will be in the

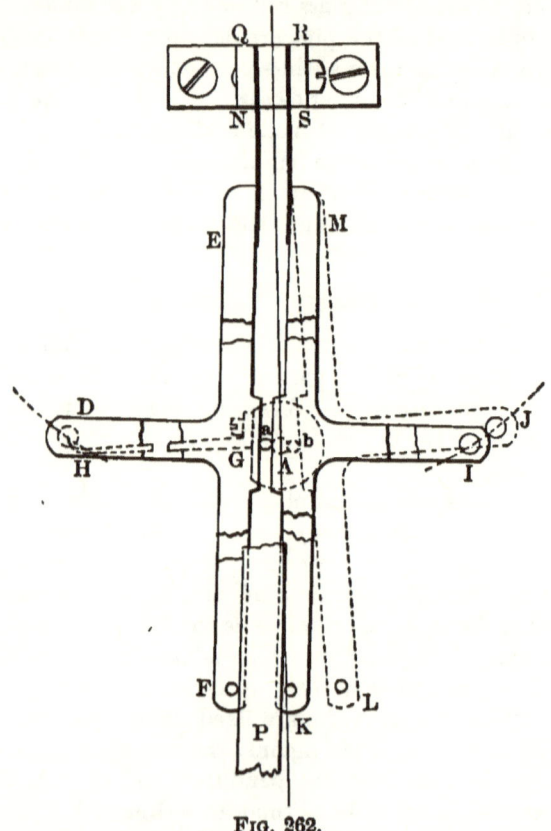

Fig. 262.

dotted position *J*, because the eccentric pin *a* has moved with *H* to the dotted position *b* and thrown the gravity piece *IKM* into the position *JLM*. This detent pin *J* detains the arm *GH* till the pendulum rod, on returning from its extreme position to the left, meets the pin *L*, and carries *JLM* along with it, thus releasing the arm *H* from the detent pin *J*, when the arm, the shaft *A*, and the eccentric pin *a* make another half-turn.

Now it is readily seen that as the pendulum rod returns, it is followed by *JLM* to the position *IKM*. That is, the pendulum rod takes the gravity piece from *JLM* to the limit of movement and back to *IKM*, one operation neutralizing the other as far back as to *JLM*; but from the excess movement *JLM* to *IKM* the pendulum receives its impulse. A like impulse is received from *F* in the opposite movement of the pendulum. The eccentric pin *a*, moved by the train of wheelwork, raises the gravity piece from *K* to *L*, or lifts its center of gravity to create the impulse.

The impulses imparted to the pendulum are thus made practically equal for all time of running of the clock.

THE CYLINDER ESCAPEMENT.

The cylinder escapement, formerly much used in Swiss watches, and interesting from the standpoint of mechanism, is illustrated in

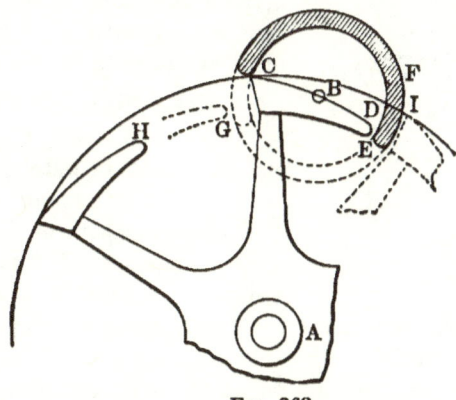

FIG. 263.

Fig. 263. At *A* is the driving staff carrying the escape-wheel with peculiar shaped teeth having inclined edges, *CD*. These teeth act upon pallets at *C* and *E*, consisting of the smoothed edges of a thin half-cylinder, *CFE*, supported to swing upon a central axis, *B*. The diameter of the inside of this cylinder is just sufficient to allow the latter to swing over the tooth *CD* with a trifle of clearance, and the distance *CH*, equal to *IG*, should be just sufficient to admit the full cylinder with a trifle of clearance.

The tooth *CD* is represented as escaping from the edge of the cylinder at *C* and soon to strike just inside the edge at *E* and to slide some distance, the cylinder turning right-handed. On reaching the limit of movement *E* returns, and on arriving at *D* the inclined edge DC of the tooth will slide against the edge E, imparting a left-handed impulse to the cylinder B. As the tooth CD escapes from E, the point H will strike just back of the edge of the cylinder at G, and slide on the outside to the limit of movement, when on returning, as the edge G passes H, the inclined tooth HI will slide along against the edge G and impart a right-handed impulse. On completing this, CD escapes at C as before. The amount of inclination of CD is arbitrary, and it may be straight or somewhat convex.

There seems to be a large amount of friction in this escapement, parts being practically the whole time in rubbing contact, which circumstance may be sufficient to explain the unspirited deportment of the escapement as observed in the watch.

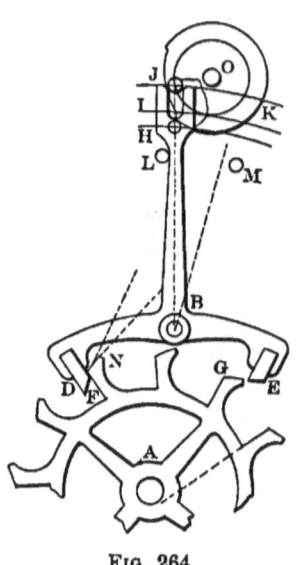

Fig. 264.

The Lever Escapement.

This escapement, Fig. 264, is the one most used in watches and known also as the "anchor escapement," "detached escapement," or "detached lever escapement," because BH is a lever, or because DEBH resembles the anchor, or because the balance parts JKO are detached and free, respectively.

This leaves the vibrating balance freest from friction of all the escapements, unless it be that for chronometers. As compared with Fig. 263, it has a very decided advantage in this respect, a fact evinced by its more lively action as observed in watches.

In this escapement A is the axis of the escape-wheel, D and E pallets of the anchor-piece DEBH swinging about an axis at B. The balance staff is at O, upon which is a collar holding a pin at J, at which the collar is cut away nearly to the pin. The lever is slotted at I and has a pin or shoulder at H. At L and M are banking pins or other provision against over-movement of the lever. In this there are really two movements, one the escapement proper between A and B, and the other a pin and slit motion between B and O. Also a lock as between the pin H and the edge of the collar, to prevent BH from moving except at the proper time.

Now supposing J to be moving toward I by the swing of the balance, when the pin J enters the slit, the pin H drops into the cutaway at J, and they will move along together toward K and become locked there upon the opposite side to that shown by the cut, and in a similar manner. When the limit of swing is reached J returns again to move HI back. Thus the balance staff and attached parts are free from contact with other pieces most of the time, since, when locked at the one side or the other, the escape-wheel teeth are locked upon the pallets as shown at F, where the angles are such as to draw D or E toward A during the lock and holding the pin H away from the edge of the collar J.

When H moves to the right, the tooth at N is unlocked and its inclined end passes the end of the pallet at F, exerting a pressure upon it and imparting an impulse to the pin J. Similar action occurs at the pallet E to give an opposite impulse to J in the reverse movement.

The pallets, or the teeth, or both should be rounded by the amount of the angle of swing of B as shown, to prevent the corners of D or N from scratching each other while in motion.

In watches, D and E are jewel pallets and J is a roll jewel.

The Duplex Escapement.

This escapement is in use in expensive and also in cheap

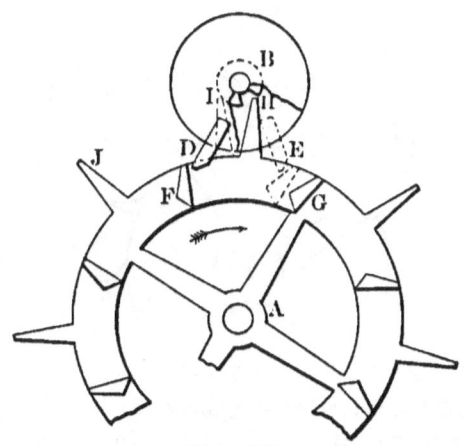

Fig. 265.

watches, and is next to the detached lever and chronometer for freedom from friction.

A, Fig. 265, is the escape-wheel, B the balance staff with a longitudinal groove at H for a short distance to admit and pass the point of the spur H, J, etc., and D is a pallet to engage the impulse tooth F.

The figure represents a spur just escaping the groove at H, and the tooth F about to engage with D, to impart to it an impulse while moving from D to the point of escape at E. Then the next spur, J, should meet the staff at I, leaving a trifle of clearance for E at G and F. The pallet moves on, to the limit of swing of balance and staff, and returns, passing to the other limit of swing, with J rubbing on the staff at I and skipping the groove H. Returning from this limit, the spur J drops into the groove H, passes from I to H and escapes, etc., as before, the escape-wheel passing one tooth at each double vibration of the balance.

The friction is chiefly the rubbing of the spur on the staff at I, and to reduce this the staff should be made as small as consistent with its strength.

This is properly called duplex, because of the two points of escape, the principal one at GE and the secondary one at H.

The Chronometer Escapement.

At A, Fig. 266, is the escape-wheel, B the balance staff, D the pallet, $LGJK$ the detent piece, G the detent, K the detent spring

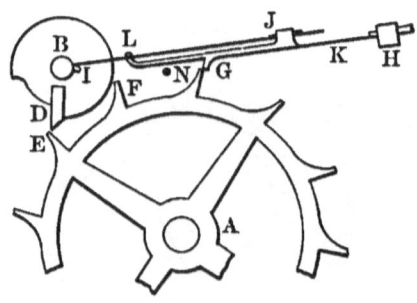

Fig. 266.

supporting LGJ and forcing it against the banking pin N, IJ the feather spring bearing against the pin L in the detent piece, and I a projection on the staff B to act upon the feather spring.

In action the projection I strikes the spring IJ resting against L, and forces LGJ away from the pin N, and the detent G away from under the tooth at G, when the tooth E strikes the pallet D, imparting an impulse. As E escapes at F, a tooth is caught by the detent G. The balance moves on to the limit of swing and returns with D just clearing F and E and with the projection I meeting the very light feather spring ILJ and flexing it to pass, when the spring flies back against the pin L. On reaching the opposite limit of swing, the balance returns, repeating the above movements in succession.

To reduce the friction to a minimum, the staff and projection at I should be as small as admissible; also the feather spring and detent spring very delicate, especially the former.

In watches and timepieces subject to abrupt displacement, the detent piece should be balanced, and perhaps pivoted, to prevent the detent from being jerked off its tooth at G.

Also if balanced on a pivot with a retaining spring to hold it up to the pin N, it should be as light as possible to prevent abrupt turns in the plane of the escapement from unlocking the detent.

PART III.

BELT GEARING.

Belt gearing includes all members in machinery concerned in transmitting motion in the manner of a belt and pulley, such as belts, bands or chains, pulleys or sprocket wheels for continuous motion; or for limited motion, where a rope, strap, or chain passes partly or several times around sectoral wheels, to which the ends are made fast, as in the windlass, or the "barrel and fusee" of chronometers, English watches, etc.

CHAPTER XXI.

VELOCITY-RATIO VARYING.

THE GENERAL CASE.

The Velocity-ratio.

In Fig. 267 take the irregular rounded pieces shown as fitted to swing about axes at A and B, with a flexible connector DE passed over and beyond those points, and made fast to the rounded bodies, so as to admit of motion to some extent by A pulling DE and driving B.

In a small displacement where D moves to G, E will move to J,

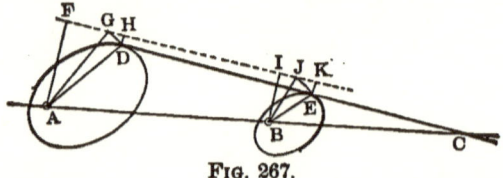

Fig. 267.

and regarding the flexible connector as inextensible in length, JK

BELT GEARING. 237

will equal GH. Also the triangle HGD is similar to FAD, and KJE similar to IBE.

If V and v be the angular velocities of A and B, we have

$$V \cdot AD = DG, \qquad v \cdot BE = EJ,$$

$$\frac{DG}{HG} = \frac{AD}{AF}, \qquad \frac{EJ}{JK} = \frac{BE}{BI},$$

and velocity-ratio $= \dfrac{V}{v} = \dfrac{DG}{EJ} \cdot \dfrac{BE}{AD} = \dfrac{AD}{AF} \cdot HG \cdot \dfrac{BI}{EB} \cdot \dfrac{1}{JK} \dfrac{BE}{AD}$

$$= \frac{BI}{AF} = \frac{BC}{AC},$$

which shows that the velocity-ratio equals the inverse ratio of the segments of the line of centers as formed by the intersection of the prolonged line of centers and connector.

The line of the connector DE is here the *line of action*, and the velocity-ratio is the same as found in all previous cases, viz., equal the inverse ratio of the segments of the line of centers, counting from the centers of motion to the intersection with the line of action.

The same is true if the connector cuts the line of centers between A and B, in which case the directions of motion are contrary, while in the outside intersection they concur.

For velocity-ratio constant the point C must be stationary, a condition readily secured for circular pulleys; also, it can be realized for a relatively long distance between centers for chain and sprocket wheels even with the latter non-circular, if symmetrical and with axes of symmetry at right angles, and for velocity-ratio equal to 1, 2, 4, etc. For the value 2, the larger wheel may be nearly a square, and the smaller nearly a duangle. But, as no advantage for this combination over circular pulleys is conceivable, they will not be further studied here.

But non-circular pulleys for velocity-ratio variable have been used with advantage.

LAW OF PERPENDICULARS GIVEN TO FIND THE PULLEY.

In Fig. 268 take A for the axis of the non-circular pulley and the periphery e, f, g, h, to be found when the line of action Dg, or of the connector, has a known perpendicular distance, Ag, from A for each 1/8 turn of A.

238   PRINCIPLES OF MECHANISM.

Lay off the several perpendiculars as radial distances, Aa, Ab, Ac, Ad, Ae, etc., and draw circles or otherwise transfer them to the several radii, marking the several angles through which A turns for the corresponding perpendiculars, giving points e, f, g, h, i. Through these several points draw the perpendicular line of the connector, as at e the line perpendicular to Ae, at f the perpendicular to Af, etc., to represent the position of the line of action for the respective radii.

Then draw the curve of the pulley tangent to these lines of

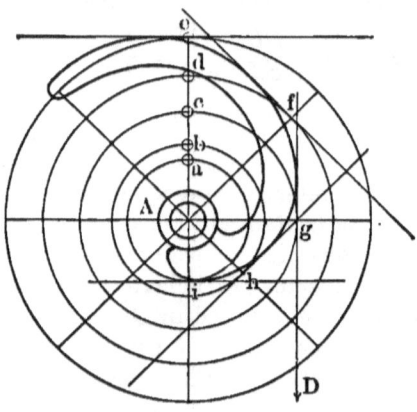

Fig. 268.

action. The pulley outline does not pass through the several points h, g, f, etc., but puts the line of action at the proper perpendicular distance from A.

Some examples may serve to illustrate.

VELOCITY-RATIO VARIABLE.

DIRECTIONAL RELATION CONSTANT.

1st. Example of the Equalizer of the Gas-meter Prover.

The meter prover consists of a hollow cylinder of some 20 cu. ft. capacity, to be raised and lowered in water for shifting air through any gas meter for testing it. The more the cylinder shell is lowered into the water, the greater the buoyancy, this being counteracted by a weight suspended from the periphery of an eccentric pulley.

To determine this pulley, let DE, Fig. 269, represent the descent of the cylinder shell into water, the surface of which is at Ee, equal steps of submergence being noted at points d, c, b, a.

BELT-GEARING.

This cylinder is suspended from a circular pulley, AF, of such size as to make a 3/4 turn for the range DE, upon the axis of which is mounted the eccentric pulley, e, f, g, h, i, of a like 3/4 turn, to which a weight, W, is suspended.

By trial or otherwise find the length of the perpendicular Ae, at the end of which the weight W will just balance DE in the highest position. Similarly find the perpendicular distance, Aa, at which W will just balance the submerged cylinder at the lowest position, or with D and a down to the water surface, Ee.

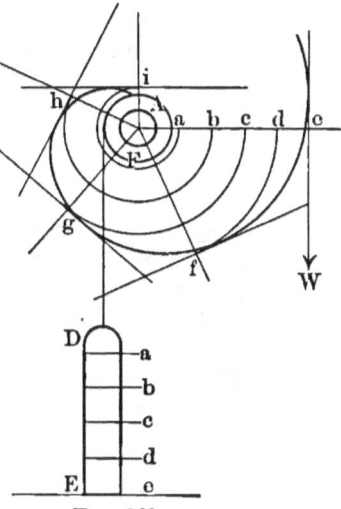

FIG. 269.

Then divide $ae = DE$ into the same number of equal parts by points b, c, d, etc., as there are equal angle divisions in the 3/4 turn e, g, i. Then make $Af = Ad$, $Ag = Ac$, $Ah = Ab$, etc., and through these points draw lines of action or perpendiculars to Af, Ag, etc. Now draw a curve as shown tangent to all these last-named perpendiculars for the linear profile of the eccentric pulley required.

The parts being made and mounted as thus determined, there should be perfect balance between W and DE in any position.

2d. Example of a Draw-bridge Equalizer.

In Mehan's "Civil Engineering" is illustrated a draw-bridge, Fig. 270, having a rope or chain with one end attached at a to the draw-bridge DE, while the other end winds upon a cylindric drum A, on the axis of which is an eccentric pulley, Fig. 271, upon which winds a rope or chain, to the lower end of which is attached a weight.

Here the weight of the bridge is constant but causes a variable tension upon the rope Aa as the bridge is opened, which is to be equalized by the eccentric pulley and weight. Observing that the weight of the bridge moves in a circle a, b, c, about D, the rope tension for the points a, b, c, etc., will be

$$\text{tension} = W\frac{Da}{Df}, \quad W\frac{Dk}{Dg}, \quad W\frac{Dl}{Dh}, \quad \text{etc.,}$$

which values are to be found and laid off on the line Aj, Fig. 271, giving points j, o, p, q, etc., for perpendiculars from A upon the line of action. Then draw an involute $astu$ to the circle A, Fig. 270, and take the distances bs, ct, du, etc., and lay off on the circle

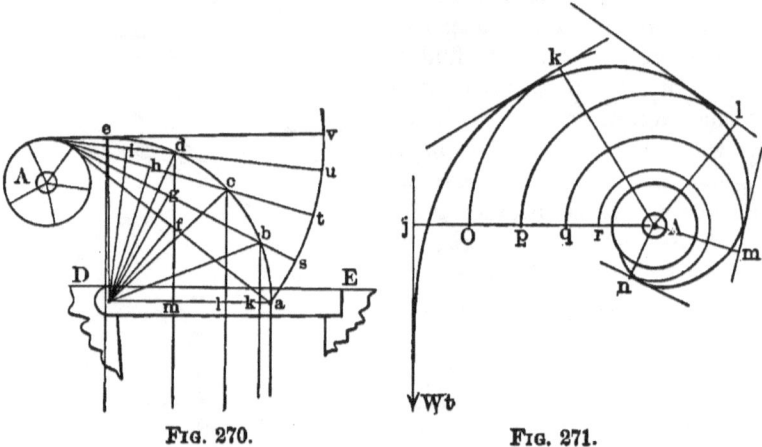

Fig. 270.   Fig. 271.

A of the drum, Fig. 271, giving angles jAk, jAl, jAm, etc., and draw the radial lines Aj, Ak, Al, Am, etc. Now lay off Ao on Ak, Ap on Al, etc., and draw perpendiculars to these radial lines through the points j, k, l, etc., and tangent to them the eccentric or spiral pulley required, as shown.

3d. The Barrel and Fusee.

In this the tension on the cord or chain, due to the torsional action of the spring in the barrel, is to be determined for equal angular positions of the barrel through the entire range of its motion. Starting with the spring slack or "run-down," find the tension when just fairly started to wind up, as for the chain or cord tangent to the barrel at D, Fig. 272. Then pull E up to D and note the tension; then F to D, G to D, etc., noting the tension for each. These will increase as the barrel is thus wound up or forced around against the action of the spring.

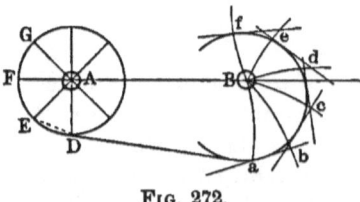

Fig. 272.

Through the center of the fusee at B draw an involute Ba to the barrel A. Then the distance from D to the involute Ba, following the cord or chain, is always the same, whether the tangent point is at D or farther toward B.

Then make the radial distances Ba, Bb, Bc inversely as the tensions as above determined, Ba for tension at D, Bb for tension when E is at D, Bc for F at D, etc.; and make the distances ab, bc, cd, etc., equal the distances DE, EF, FG, etc. Through B and these points b, c, d, etc., copy the involute Ba, as shown, and at these points draw normals to the involutes. Then, tangent to these normals, draw in the curve of the fusee.

In this construction, the smaller the angles DAE the more accurate will be the result until the limit is reached where the inaccuracies of graphical work predominate. For relatively large angles, DAE, etc., it will be advisable to make the distances ab, bc, cd, etc., which are chords to the arcs, equal the straight dotted line or chord DE. Fig. 272 is made as if aD were to act by compression. To change the figure for tension in aD, place b, c, d, etc., to the left of Ba.

The "snail" used on spinning mules is like a double fusee, winding a cord from small to large and then back to small radii again, thus varying the velocity of the cord taken up; or of the snail, when the cord is made fast at the remote end.

In deep mines, where heavy wire ropes are used for hoisting, the winding drums may be made conoidal to compensate for the varying load due to the varied weight of rope run out as the hoisting cage is let down, and *vice versa*. If the length of rope run out per revolution were constant, the drum would be a cone; but as more rope is let off per revolution where the drum is larger, its shape will be a concave conoid.

4th. Non-Circular Pulley for the Rifling Machine.

There was in use in the rifle factory of Windsor, Vt., fifty years ago, a rifling machine in which was employed a belt and non-circular pulley connection, for the purpose of imparting to the rifle groove-cutting tool, held in a rod, an approximately uniform motion forward and back as the tool and rod traversed the rifle barrel. The rod was driven by a crank and pitman, on the crank shaft of which was an elliptic-shaped pulley similar to Fig. 275 connected by belt with a circular pulley above. For the slow motion here employed this belt and pulley combination worked satisfactorily.

To determine the correct form of non-circular pulley on this crank shaft to give a uniform motion to the slide, consider the pit-

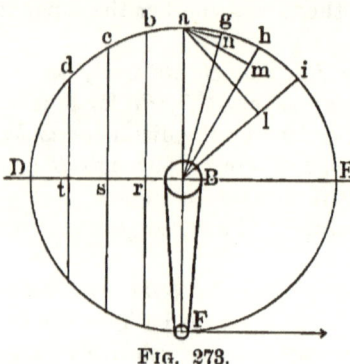

FIG. 273.

man of infinite length, when the projections of the several positions of the crank pin a, b, c, g, h, i, etc., upon the line DE, Fig. 273, will divide the latter into equal parts, r, s, t, etc., representing the uniform motion of the slide, as if the crank pin F of the crank BF were moving through those points. Drawing Bg, Bh, etc., we obtain the angles the crank must pass in equal time. On these lay off the respective perpendiculars br, cs, etc., giving Bn, Bm, Bl, etc. At the ends of these draw perpendiculars as in Fig. 268. These will all pass through the point a, as na, ma, la, etc., and drawing in

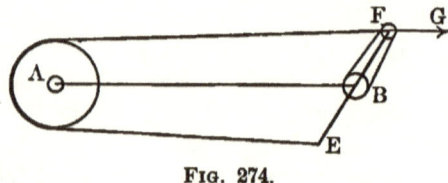

FIG. 274.

the pulley curve tangent to these perpendiculars gives simply the point a, which shows that the resulting non-circular pulley is merely a flat bar, EF, Fig. 274. Though this seems at first unreasonable, it is correct, since the belt from the uniformly-moving circular driving pulley A, going to the end F of the pulley FE, would move

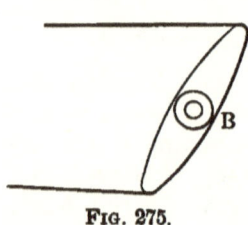

FIG. 275.

F uniformly in the direction of FG, thus giving the crank pin at F the motion required. Hence the pulley EF should be a flat bar with a very rapid motion when the crank pin is on the line of centers. This pulley would work with a suitable distance between centers without a serious variation of tension of belt.

At the dead centers of the crank, the jerk of speed the theory would give may be avoided by arbitarily widening the pulley somewhat as in Fig. 275. For moderate speed, as for a rifling machine,

BELT GEARING.

this belt and pulley combination would doubtless be thoroughly practical.

DIRECTIONAL RELATION CHANGING OR CONSTANT.

Prof. Willis has devised a movement answering to directional relation and velocity-ratio varying, called "*cam-shaped pulley and tightener pulley*," where any non-circular pulley may be mounted on any center and mated with a fixed circular pulley and a tightener pulley on a lever. The eccentric pulley should be the driver to avoid slipping of belt. (See *Willis*, page 201.)

Example of a Treadle.

One application of a wrapping connector with directional relation changing is found in the foot-lathe treadle, where an eccentric pulley, *E*, is put on the driving shaft, and a centrally mounted pulley, *F*, on the treadle bar, *BD*, and the two connected by a belt, the arrangement serving to avoid making a crank in the driving shaft, as in Fig. 276. The action is evidently the same as if a pitman were used to connect the centers *E* and *F*.

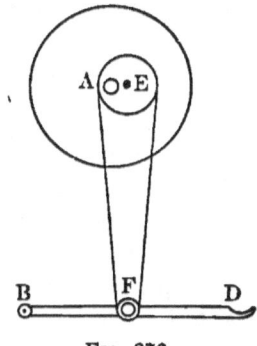

Fig. 276.

It is not necessary that the pulley *E* be an eccentric, but it may be an ellipse, with *A* in one focus, or a flat bar clamped at one end to the shaft *A*.

In case of the ellipse of the same length as the diameter of the eccentric, and with *A* at the same distance from the center, the full stroke of the treadle bar will be the same for both, but the ellipse will lower the treadle more slowly at first and more rapidly in the last part of the stroke. If *A* be at the center of the ellipse and the latter considerably longer than the diameter of *F*, the treadle will have two short double strokes per revolution instead of one.

CHAPTER XXII.

DIRECTIONAL RELATION CONSTANT.

VELOCITY-RATIO CONSTANT.

HERE the pulleys are both circular, of any relative dimensions, at any distance apart, and with belts open or crossed as in Fig. 277.

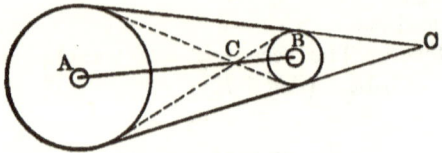

FIG. 277.

The velocity-ratio is always equal to $\dfrac{BC}{AC}$ whether the belt is open or crossed, with C outside or between centers. When C is outside, the pulleys turn the same way, and when between, they turn in opposite directions.

The exact velocity-ratio is difficult to obtain, for two reasons: first, the thickness of the belt adds somewhat to the practical diameter of the pulley, and relatively more for the small than large one; and second, the elastic yielding of the belt, contracting as it goes around the driving pulley, and expanding as it goes around the driven pulley, causing a "slip" of belt which increases with the driving load and slackness of belt. This slip keeps the pulleys bright.

The Belt.

The belt may be a strap of leather, or of woven stuff sometimes filled with rubber; or of leather links on wire joint pins built up to any desired width; or a strap with triangular blocks attached to run in a V-grooved pulley; or a round belt formed of thick leather, first cut into square strips and then rounded by drawing through a die when small, as for driving sewing machines; or when larger, from ⅜ of an inch diameter up, it is made of a flat strip twisted and pressed; or a rope of hemp or wire. All but the flat belts run in V grooves.

Chains running over sprocket wheels now constitute a most important connector.

Retaining the Belt on the Pulley.

A V groove of sufficient depth will naturally retain the round belt or rope. But for the flat belts, the pulley is made "high center," since a flat belt has a tendency to climb to the highest part. This is due to the edgewise stiffness of the belt, giving it a tendency to climb a cone of moderate taper.

Thus, in Fig. 278, take OAB as a cone cut at the lower element and developed by rolling out to DE till coincident with the tangent

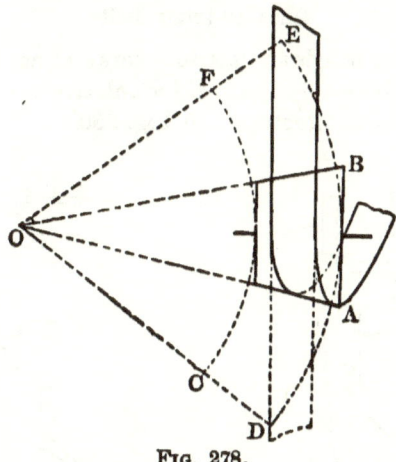

FIG. 278.

plane ODE. Placing a strip of belt, DE, upon it, we see that it runs off the developed cone surface at D and E, though at the middle point it is parallel to the base.

Redeveloping the portion of conic surface CD, and the belt with it, the latter is seen to run off the base of the cone, getting more and more inclined the farther it goes.

246   PRINCIPLES OF MECHANISM.

Hence it appears that a flat belt running on a pulley with a high center will climb to the highest point.

Advantage may be taken of this fact to carry a belt, not very wide, at quite an angle to the shafts, to avoid an obstacle, A, as in Fig. 279. It is here only necessary to make the pulleys quite convex and a little larger at one end.

Crossed Belts.

In a crossed belt the latter must have a twist of 180 degrees between pulleys to keep the same side of belt to pulley. Thus the edge elements of belt must be appreciably longer than the central ones, by the ratio of the hypothenuse of a triangle to its base, when the latter is the length of the free part of belt, and the altitude of the triangle the half-circumference of a circle whose diameter is the belt width. Hence it is not feasible to run a wide belt crossed for a comparatively short line of centers.

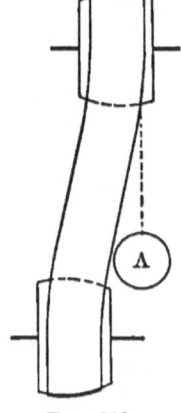

FIG. 279.

Quarter-twist Belts.

A quarter-twist belt is subject to severe strains, due to distortion, as well as the crossed belt, more for relatively large pulleys and less for small ones, as illustrated in Fig. 280.

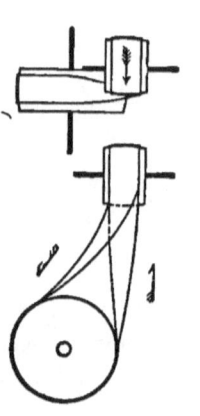 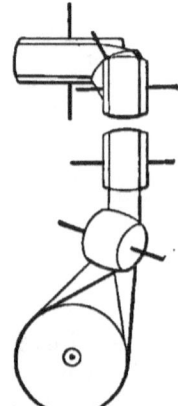

FIG. 280.   FIG. 281.

Without a guide pulley, the belt will immediately **run off the** pulleys if the latter are turned backwards.

By arranging the pulleys as in Fig. 281, and adding the guide pulley to hold the guyed part of the belt over near to the straight portion, the pulleys may run either way without throwing the belt off and with less strain on it. The guide pulley, however, must be placed in an awkward position, its central transverse plane to coincide with the center lines of the belt above and below that pulley.

Any Position of Pulleys.

The driving and driven pulleys may be placed in any possible relative positions and connected by a single belt, if four or less guide pulleys be provided, four for the comparatively simple case where the pulleys are on parallel shafts and not in the same plane. For more awkward positions it will usually happen that tangents to the driving and driven pulleys can be made to intersect, at each point of which a single guide pulley may be located, with meridian plane coincident with the plane of the proper tangents.

Cone Pulleys.

Pulleys in a series of steps, as employed on lathes and their countershafts, by which the speed of the lathe may be changed from one constant speed to a number of others by shifting the belt are called cone pulleys or stepped cone pulleys. Not only lathes but a large variety of other machines require these cones, so that their correct construction is important.

A common practice is to make these steps equal on one of the cones, when they will require to be equal on the other for a crossed belt, but for an open belt not for uniform tightness of belt.

A little inquiry will satisfy the seeker after truth that the cone diameters should be such as to place the steps of speed in geometrical progression, that is, the relation of any one speed to the next should be the same as that to the next; or the ratio of any two adjacent speeds should be constant throughout, as has been correctly maintained by Professor John E. Sweet.

For instance, in a "back-geared" lathe, the ratio of speed for the first two sizes should be the same as for the last two sizes of the cone, in order that in back gear the ratio of speed may be preserved. Thus, for simple ratios, suppose the cones give the geometric series of speeds as 1, 2, 4, 8. Then the back gear should continue this series as 16, 32, 64, 128; the ratio of any two adjacent speeds being 2. That due to going into the back gear should still be the same. Otherwise, suppose that the series of speeds were assumed as 1, 2, 3, 4; when the back gear will give 5, 10, 15, 20, if the gearing is such as to give the first figure, 5. Then the ratios of adjacent speeds will give the series of ratios

$$2, 1.5, 1.333, 1.25, 2, 1.5, 1.333,$$

a quite irregular set of figures. Granting that they are correct up through the cone to back gear, the values of the ratios decreasing gradually from 2 to 1.25. Then, instead of continuing on a decrease, there is a sudden jump to 2 again, after which a second decline—which is clearly irrational.

It is also important that the gearing of the back gear be proper to this rational geometric series of Prof. Sweet as well as the cones, that the ratio of adjacent speeds throughout may be a constant.

A convenient way of realizing in a drawing this geometric series of speeds in laying out a pair of cone pulleys is shown in Fig. 282, where a length AD, measured by some scale, represents the revolu-

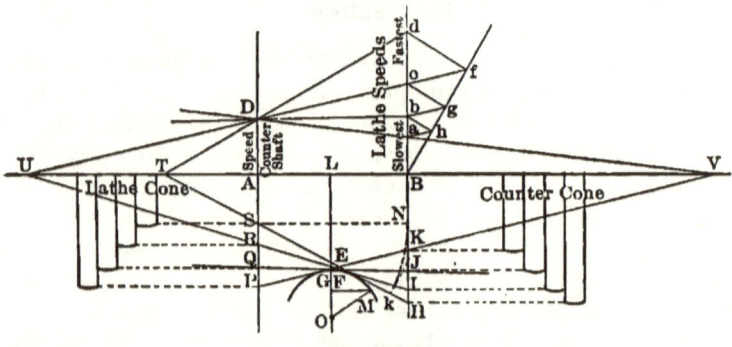

FIG. 282.

tions per minute of the countershaft; Ba, Bb, Bc, etc., by the same scale, the speeds of the lathe or other cone; BH, BI, BJ, etc., the radii of the countershaft cone; AP, AQ, AR, etc., the radii of the lathe cone; and AB the distance between the axes of the cones. This may be correctly drawn as follows:

From B draw a line Bf at any convenient angle. Draw lines ah and hb, then bg, gc, cf, etc., parallel to them respectively, when the distances Ba, Bb, Bc, etc., will be in geometrical progression and will represent the geometric series of speeds of the lathe cones, Ba being the slowest, and the others Bb, Bc, etc., being made by trial to agree with the series required.

CIRCULAR BELT GEARING.

Draw lines Da, Db, Dc, etc., and extend them to intersect the line AB, also extended. From these points of intersection draw lines TSH, URI, VKP, etc., tangent to the circle GM, drawing the first line through S so that AS may measure the desired smallest radius of the lathe cone; or, if preferred, the line VKP may be drawn instead, making AP the desired largest radius. When the first line is drawn, strike in, tangent to it, the circle GM, from a center O on a line midway between AP and BH. The radius of this circle is

$$OM = \frac{AB}{\pi} \text{ nearly,}$$

and is to be calculated. The remaining lines, URI, QJ, etc., are drawn tangent to this circle, giving the radii of the lathe cone AP, AQ, AR, etc., and radii of the countershaft cone BH, BI, BJ, etc.; and the cones may be drawn as shown.

If a line as Db does not intersect AB on the drawing board, we must make $Bb : BJ :: AD : AQ$, which may be done on another diagram.

The diagram shows that

$$AD : Ba :: AP : BK;$$

that is, the speed of the countershaft is to the slowest speed of the lathe cone as the largest radius of the lathe cone is to the smallest radius of the counter cone, as it should be to accord with the geometric series.

In a diagram for a crossed belt the lines all meet in one point near GE, while for the usual case of open belts the positions of the lines must be "doctored" by being drawn tangent to the circle GM, which shifting of the lines, however, does not interfere with the velocity-ratio, since the ratio AS to BH will always be the same when the line is drawn through the point T.

To determine the radius OM, let ΔR and Δr, Fig. 283, stand for the corrections of the radii R and r to account for foreshortening due to the inclination of the belt, which amount call Δl. Then as TW is the belt inclination, half the foreshortening due to it is

$$kW = XW - SN = 1/2\ \Delta l \text{ nearly, and } = \frac{\pi}{2}(\Delta R + \Delta r),$$

whence

$$\Delta l = \frac{2\pi(\Delta R + \Delta r)}{2} = 2\pi GE.$$

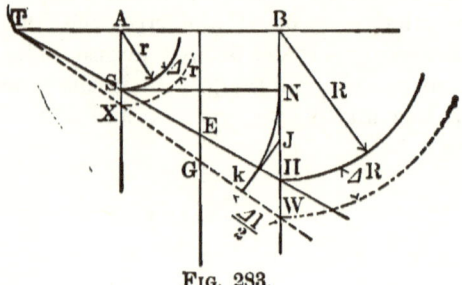

Fig. 283.

Also, by Fig. 283,

$$\frac{\Delta l}{2} : \frac{R-r}{2} :: R-r : AB \text{ nearly,}$$

giving $\quad (R-r)^2 = \Delta l \cdot AB = 2\pi GE \cdot AB.$

Again, Fig. 282,

$$2GE : FM :: R-r : AB \text{ nearly,}$$

whence

$$R - r = \frac{AB \cdot 2GE}{FM}.$$

Eliminating $R - r$, we get

$$\frac{\overline{FM}^2}{2GE} = \frac{AB}{\pi}.$$

But $\quad OM : FM :: FM : 2GE \text{ nearly,}$

giving

$$OM = \frac{\overline{FM}^2}{2GE} = \frac{AB}{\pi}$$

the required radius.

This radius, or rather the position of the point O, Fig. 282, differs from that adopted by Mr. C. A. Smith, Trans. A. S. M. E.,

Vol. X., p. 269, where $LO = 0.314\, AB$. The latter may be used in constructing Fig. 282, if preferred.

In the extreme example of $AB = 48''$, $AS = 1''$, $BH = 17.284''$, while an intermediate pair of sizes were 10'', the difference of belt length, as given by Fig. 282, and the carefully calculated value, differed by only about 0.2''.

Reuleaux, in *The Constructor*, p. 189, gives an interesting diagram for determining cone pulleys; also Prof. J. F. Klein, in *Machine Design*, diagrams and tables; but in none of these do we find the very important consideration of the geometrical series of speeds.

In practice the drawing for Fig. 282 may take an uncomfortable length. It may be shortened to the extent that the angle BTH does not much exceed 30 or 40 degrees, by taking for the actual distance AB a fictitious length ab, and using for the radius OM the value

$$OM' = \frac{(ab)^2}{\pi AB}.$$

Actual cones have been employed where a variation of speed is desired while running, the belt being shifted when desired. It is usually difficult to keep the belt running satisfactorily, unless the cones are unduly long.

The Evans friction cones is a good arrangement where cones are required, in which the cones are placed with large and small ends opposite, and a hoop of belting around one cone, somewhat larger than the large end, arranged with a guide, which hoop makes the bearing point between the cones. By shifting the guide the hoop is shifted and the speed changed.

Rope Transmission, Rope Belting, Etc.

The term rope transmission was first applied to rope belting used for transmitting power over relatively long distances, but recently it has come to be employed as a substitute for leather belting. For transmission wire rope has largely been employed without a "take-up" where the sag of the rope for stretches of several hundred feet will vary, compensating for temperature and wear, and still maintain sufficient tension for service. With this tension and a speed of the rope varying from 3000 to 5000 ft. a minute, a large amount of power will be transmitted with only a half-turn of rope over the pulley for frictional contact.

But for shorter spans, where hemp or cotton rope is used, provision is made, first, for "take-up" for maintaining constant, or at least sufficient, tension; and second, for drawing the rope as it stretches. In cable railways, where the working tension of the rope is carried to a very high value with comparatively low speeds, the take-up and increased frictional contact with the driving drum are matters of the utmost practical importance. In the slower speed arrangements the rope is to be consumed by a higher working tension, while in the higher speeds the rope is worn out by a high working velocity, causing frequent flexural stresses and abrading contacts at the pulleys.

In Rope Transmissions there are two leading systems: first, where the one loop of rope reaches the entire stretch of the system, with a driving pulley at one end and a driven one at the other, guide pulleys of less diameter being introduced at points between to carry the rope, the whole layout of rope being in one vertical plane; and second, where one loop of rope passes over a driving and driven pulley only, with no guide pulleys, beyond which is a second loop of rope, likewise mounted and driven from the first, and beyond which last is a third loop of rope similarly mounted and driven from the second, and so on for as many bents as desired; there being at each intermediate station point either two pulleys made fast on one shaft or a double pulley. This system must be arranged in one vertical plane, but, as in the first, may pass over hills and valleys.

The driving and driven pulleys are of large size; first, because of the high speed required, and second, to prevent a too severe flexing of the ropes. The bottoms of the grooves in the pulleys are fitted with wood, hard rubber, leather, or like material, to increase the friction of contact with the rope.

A horizontal angle may be turned in either of these systems in at least two ways: first, by introducing bevel gears at any station; second, by passing a rope vertically downward or upward from one pulley and immediately upon another set in a vertical plane making any desired angle of deflection with the first plane.

In the above, wire rope is usually employed.

In Rope Belting several half-turns are made around the driving and driven pulleys to produce the necessary frictional contact. This is done differently for long than for short stretches between power pulleys.

ROPE TRANSMISSION. 253.

First. **For a Short Stretch**, Fig. 284 illustrates a mode of duplicating the passes of rope from driving to driven pulleys. The rope passes from the take-up or tension pulley D to the first groove of

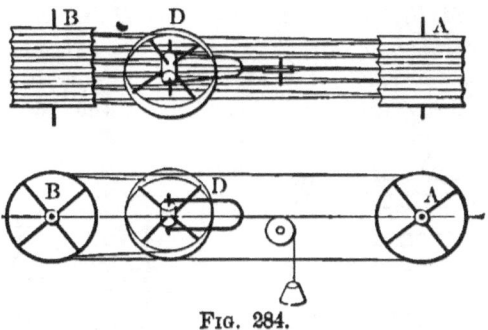

FIG. 284.

B. Passing half around B it runs to the first groove of A. Passing half around A it goes over to the second groove of B and a half around, when it goes to the second groove of A, and so on for all the grooves of A and B till when it leaves the last groove of B it passes to and half around the take-up pulley D, when it goes to the first groove of B again and repeats the circuit. One piece of rope spliced into a single loop makes the entire circuit.

The take-up pulley D is mounted upon a slide, and has a weight to produce the desired tension and take up the slack due to varying length of rope.

If the pulley grooves of A are all of one diameter, likewise of B though it may differ from that of A, the ropes in the working tension side will all have the same tension, and those on the slack tension side will be in equal tension with each other, but less than that on the working tension side, but not if the grooves differ in diameter on either one of the wheels.

The pulley D may be placed outside of A and B, as if the rope passed from B over beyond A and then back to B again, but with no advantage.

Second. **For a Longer Stretch**, the rope belt may pass over a counter pulley at each end of the system, as in Fig. 285.

Here the rope between A and B is in higher tension than in Fig. 284, it being due to the cumulative action of all the half-turns of rope contact with the driving pulley.

One point to be noticed here is the fact that the rope is varying its tension, or seeks to, from the first groove of A to the last, and consequently varying its length. On this account the pulley A should have its grooves varied in diameter from one end to the other, and likewise for the counter pulley D. The same should be done at the end B. The counter pulleys D and E, however, may consist of separated sheaves loose upon the axis and of one size. The amount the grooves of A and B are to vary in diameter will depend upon the unit of elastic yielding of the belt, and the total

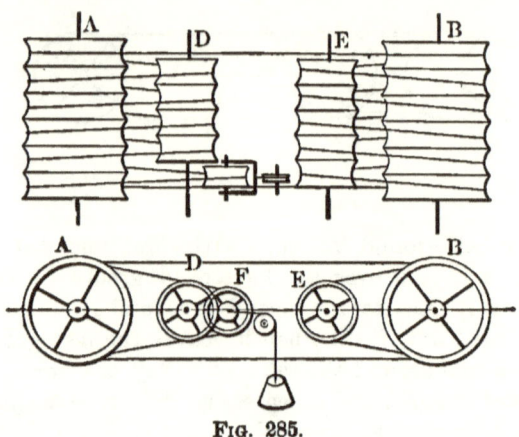

Fig. 285.

variation of tension between the going and returning sides. The drop in tension from one end to the other of A or of B should be uniform per groove, and likewise the diameter; while the end diameters should differ by the amount an equal length of belt will vary as due to the total variation of tension.

The counter pulleys D and E may both be stationary, with a single turn of belt going over to a single-groove tension pulley F. Also, D and E may be geared up with A or B, and become power pulleys.

Third. **For Cable Railways, Haulage Lines,** etc., the pulley and counter pulley A and D are connected and driven by power, while at B there is only a single sheave for returning the rope.

Ingenious compensating devices are in use to provide for the variation in rope length while passing over the series of grooves on the connected driving and counter pulleys.

For instance, let G and H, Fig. 286, represent two pulleys of equal size and number of pulley grooves, tapered from H toward G to partially compensate for varied rope length between H and

G. The two pulleys are loose on the shaft *F*, but driven by a bevel gear at *IL* and *JK*, mounted upon axles at right angles to the shaft *F*, and made fast upon it. These gears mesh into a bevel gear *IJ* fast in the pulley *H* at *IJ*, and into a bevel gear *LK* fast in the pulley *G* at *LK*. Thus mounted, if *H* is turned one way, *G* will be rotated in the opposite direction, by reason of the bevel gearing interposed.

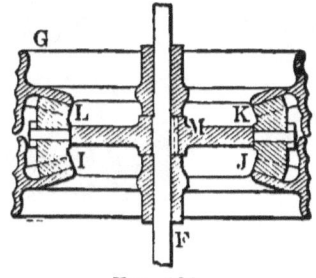

FIG. 286.

A second system like Fig. 286 may be prepared and the two connected by gearing to serve as *A* and *D* in Fig. 285. When rigged with rope and started into service the varied diameter of *H* will partly compensate for varied tension and length across *H*, and likewise across *G*; but between *H* and *G* the gearing will permit a perfect compensation, by revolving slowly in opposite directions. Then, when the rope is working under a greater tension than the taper of pulleys *H* and *G* provides for, they will move one way on the shaft *F*, and the opposite way for a less tension than the taper of pulleys provides for. Thus the creep of rope at contact with pulleys is reduced by the splitting of the pulleys from one into the two, *G* and *H*.

The Chain and Sprocket Wheel.

A wrapping connector, connecting wheels with an exact velocity-ratio, is found in the chain so formed that it will engage the teeth of a spur or sprocket wheel, as in the familiar example found on the safety bicycle, for a connection between pedal axis and driving wheel.

The chain is made with various forms of link, one of the earlier ones being shown in Fig. 287, where the links are punched out of sheet metal and pinned by rivets, while a more modern one is shown in Fig. 288, all links being of the same form, sometimes cast and sometimes drop-forged. The latter possess important advantages in having the axial pin solid with the rest of the link, and in having each piece of the chain like every other, all hooked together. A link, turned into the position shown at *D*, may be removed from the hook and more or less links added, while for working positions the shoulders *a* and *b* prevent unhooking. It is thus easy to shorten a chain when worn and slack, and by the amount of one link length; while for Fig. 287 a smith is required, and the least that can be removed or added is two link lengths.

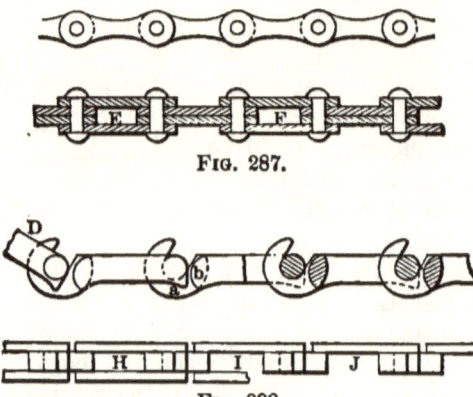

Fig. 287.

Fig. 288.

The Teeth of a Sprocket Wheel engage at E, F, etc., or H, I, J, etc., Figs. 288 and 289, as the case may be; the proper forms of

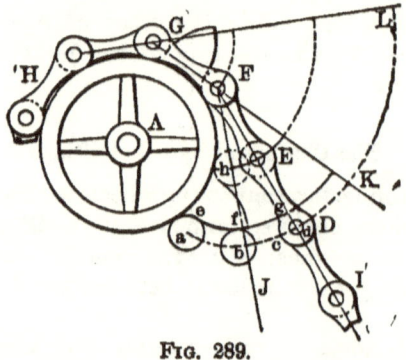

Fig. 289.

the teeth being determined as shown in Fig. 289.

Take HGF a portion of the chain on the wheel, while at F it runs off in the tangent FED.

Now as FD is wound upon the wheel, the center point E describes a circle arc about F till the link head at E strikes the rim of the wheel, when the links EDI will swing about h as a center till D strikes the rim of the wheel, and so on. Then we will find that the arc from a to the straight line FJ is circular about h as a center; from F to GFK it is a circle arc with F as a center; and from K to L it is a circle arc about G as a center, etc. Then a series of circle arcs ef, fg, parallel to ab, bd and at a distance dg from it, will be proper for a sprocket tooth profile for the circular wheel A. This curve copied around will give all the teeth.

For non-circular sprocket wheels, as in the cases of the elliptic ones used on some bicycles, the sprocket-tooth outlines, to follow theory exactly, should each be determined in the manner of Fig. 289.

Non-circular sprocket wheels have a variation of velocity-ratio the law of which is generally simplest when one of the pair is circular; but both may be non-circular and in any ratio of sizes, provided the wheels both have symmetrical axes at right angles to each other, observing that the pair of wheels should be so far apart that the inclination of the chain to the line of centers does not become so excessive as to vary the tension of the chain unduly.

Practical Application of Chains.

Chains are often used to retain a mathematical relation between a pair of axes where other wrapping connectors would not answer.

They have been tried for the transmission of power, but experience shows that they must run so very slowly to prevent noise and shocks that they have been abandoned for this purpose.

Hooks and projections have been formed on the links to which may be attached conveyor bars, boards, buckets, scoops, etc., when, in stretches of considerable length, sometimes several hundred feet, running vertically or on inclined tracks, grain, sand bags, refuse, etc., may be carried in a continuous current. Thus excavators have been operated, and coal-cutting machines, where cutting tools are made fast to the links.

PART IV.

LINK-WORK.

This term is applied to such machinery as consists of rods, cranks, levers, bars, etc., jointed together, either for axes parallel or meeting, or crossing and not meeting.

CHAPTER XXIII.

THE GENERAL CASE.

The Velocity-Ratio.

Take A and B as fixed centers of motion, AD the driving crank or lever, BE the driven crank, and DE the connecting-rod bar or link. As AD turns about its center A, the rod DE compels BE to turn also; Fig. 290.

To determine the velocity-ratio compare with Fig. 267, in which

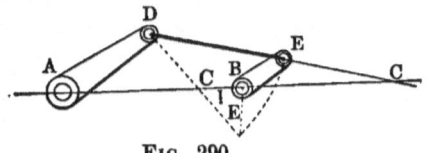

Fig. 290.

AD, DE, and BE may replace the lines of like lettering of Fig. 290. Hence for the latter the

$$\text{velocity-ratio} = \frac{V}{v} = \frac{BC}{AC}.$$

When the point C is beyond AB the cranks turn in the same direction, and in opposite directions when it is between.

Peculiar Features of Link-Work Mechanism.

Link-work movements are the most arbitrary in law of motion of all the combinations of mechanism, and in many cases where it would otherwise be preferred, must be abandoned on this account. It is the "lightest-running" mechanism known, the resistance being due to the slight friction of comparatively small pins in bearing holes well lubricated, and with comparatively long arms with which to turn those pins. Also, a pin rarely makes more than one complete turn in its bearing hole in a complete movement, and usually much less; while in the corresponding movement of a cam motion the roller (as in the best arrangement) makes from 6 to 12 turns on its pin, and even this is not so prejudicial as regards resistance as the rolling of the roller along the surface of the cam groove.

Link-work is consequently much more durable than other forms of mechanism, and should be adopted into machinery whenever it can be in preference to cam-work, tooth-gearing, etc.

It often happens that the principal or hardest working movements of a machine may be link-work while cam motions adapted to it may serve for other movements, and give, on the whole, a more satisfactory machine in operation than where all are cam movements.

Here, instead of assuming a law of motion and finding the link-work, as can be done in gearing, cam and belt movements, etc., we must examine a proposed link-work movement complete, when, if found unsuitable, try another, etc., till an acceptable one, if possible, is obtained.

The motion sought may be impossible, while an approximation to it may answer by conceding some unimportant idiosyncrasy of movement.

I. Axes Parallel.

Most link-work belongs to this case of axes parallel, and the treatment must largely be by examples, owing to peculiarities and want of susceptibility of reduction to broad and extended laws.

Some Examples of Link-work Movements.

Example 1.—To illustrate: An excellent needle-bar motion for a shuttle sewing-machine is obtained by link-work; and it is, without question, in one case at least, the lightest-running shuttle machine yet built, the whole machine being link-work mechanism.

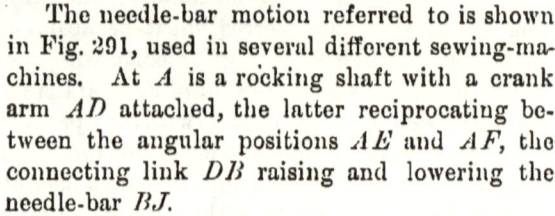

Fig. 291.

The needle-bar motion referred to is shown in Fig. 291, used in several different sewing-machines. At A is a rocking shaft with a crank arm AD attached, the latter reciprocating between the angular positions AE and AF, the connecting link DB raising and lowering the needle-bar BJ.

As D passes the line AI of crank and link straightened, the bar reaches its lowest position. As D moves on to its limit of motion at F the bar is raised an amount GH, forming the loop for the shuttle. Now it is not necessary for the needle-bar to return to G, but on tarrying a moment raised to H for the shuttle point to fairly enter the loop, it may reasonably enough continue on its upward journey. By a cam it would be given this action, but by the link-work of Fig. 291 it *must* return to the lowest point again before making its ascent.

Now if this drop by the amount HG, after the shuttle enters its loop, is considered fatal to the machine, the proposed link movement of Fig. 291 must be abandoned; otherwise it may be accepted, giving a much lighter-running and more durable combination than if the cam and pin be adopted, or such a movement as in Fig. 252.

Example 2.—Another instance is found in the Corliss valve gear, where the rocker plate swings to carry the pin D back and forth from H to F. The valve and stem B are moved by the lever BE as it swings from BE to BG. As D moves from J to F and return the valve is opened and closed, and from J to H and back it remains closed, but is compelled to be moving

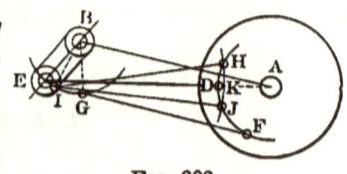

Fig. 292.

slightly, as due to the double versed-sine DK. If this movement DK is counted out of the question, the link-work combination of Fig.

292 must be rejected, otherwise accepted, as in fact it has been by hundreds of builders following in the footsteps of Geo. H. Corliss.

Example 3.—Another example is given in Fig. 293, where the link-work movement rotates a shaft B, by the arm BE, from the position BF to BE and back, during the half-turn of the main shaft that drives AD from AI to AD and back; while for the remaining half-turn of the main shaft, D moves from I to II and back, during which BF is nearly stationary, its simultaneous movement being from F to G and back. It was preferred that for this latter BF remain absolutely stationary, but the slight movement FG was counted less ob-

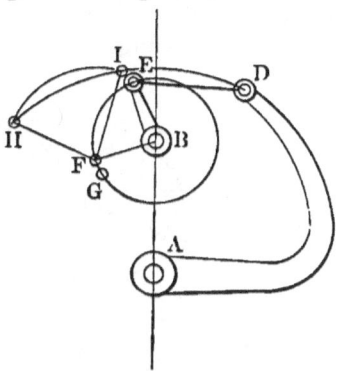

Fig. 293.

jectionable than cam-work, as compared with the linkage of Fig. 293, so that the latter was adopted.

It may be noted that the slight movement FG is less when the two arcs III are convex the same way than when convex opposite ways, and more so as A is nearer to B.

Example 4.—As another example, suppose a point is required to move from E to F, Fig. 294, and return within the sixth part

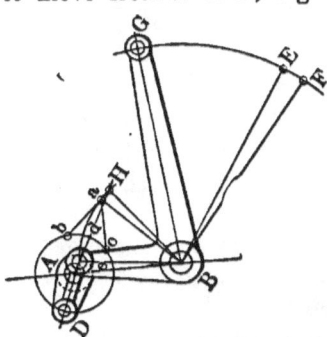

Fig. 294.

of a turn of the main shaft A. A cam may be used, by which the point is thus driven and then allowed to remain quiet at E for the 5/6 remaining turn. But if there is no particular objection to the movement of the point from E to G and return, instead of remaining stationary at E, link-work may be employed, as in Fig. 294, where A is the main shaft, AD the crank, De the pitman, and $eBG = HBF$ a bell-crank lever, the arrangement being such that while the crank pin moves from c to b the required sixth part of a turn of A, a moves to H and back, and E to F and back, thus meeting the essential conditions of the movement.

Path and Velocity of Various Points.

In the study of the motion of linkages it is sometimes necessary to determine simultaneous positions throughout the movement for all the joints, beginning with the driver which, for uniform motion, should have the points equidistant, as illustrated in Fig. 295. Here the driver *A* makes its circuit with uniform motion from 0, 1, 2, etc., to 8. Corresponding points are given like numbers throughout. The link *BE* swings about the fixed point *B*, and the end *F* of the

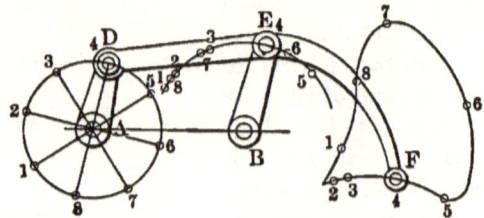

Fig. 295.

link *DEF* describes the curve *F*, 5, 6, etc., and returns to *F*, all the spaces being passed in equal times provided *A* moves over its numbered spaces in equal times.

The curve described by any point of the link *DEF* may be determined, as was done for the point *F* in the above instance. Modifications of the path *F*, 5, 6, etc., may be made by changing the position of *B*, or length *BE*, or angle *DEF*, etc.

A link may be connected at *F*, thus extending the linkage into a train, and possibly several links in succession, leading through several elementary combinations.

Sliding Blocks and Links.

Sliding parts are sometimes introduced, either for realizing the conditions of a link of infinite length, or for simplifying the mechanism, a notable example being that of the crosshead of the steam-engine.

Sometimes a link may be greatly shortened and simplified in construction, as shown in Fig. 296, where the pin *F* moves in some curve *FE*, as that described by the point *F* in Fig. 295. In fact the parts shown in Fig. 296 may be employed in continuation of those of Fig. 295 into a train, *F* moving through the points 1, 2, 3, 4, etc., in equal times, being regarded as the driver for the elementary combination of Fig. 296. The point *G* is the fixed fulcrum pin for the bell crank *HGD*, and points 1, 2, 3, etc., denoting positions for *H*, may be found as corresponding with like figures of the curve *FE*, as if that curve were the one through *F* of Fig. 295, or again, as corresponding with the points 1, 2, 3, etc., of *AD*, Fig. 295.

The block *FD* serves as a short link connecting the straight pins *F* and *D*, upon both of which the block slides to accommodate the versed sines of the curve *FE* and of the circle arc about *G*.

As these pins are always perpendicular to each other, the block *FD* is a simple single piece with the two holes at right angles to

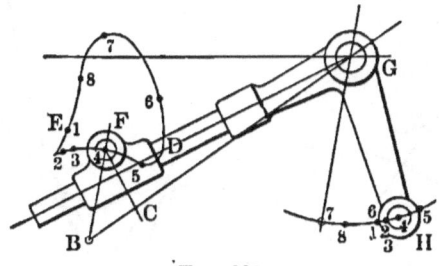

Fig. 296.

each other. In this way the motion of *DH* is to be found as in sliding contact, and on this account the movement would seem to require classification there. To determine the velocity-ratio, find the center of curvature *B*, of the curve *FE* at *F*, and draw the normal *FC* to the pin *D*, and the velocity-ratio $= \dfrac{BC}{GC}$. This becomes simple when the curve *FE* is a circle.

CHAPTER XXIV.

THE ROLLING CURVE OR NON-CIRCULAR WHEEL EQUIVALENT FOR LINK-WORK.—GABS AND PINS.

For every elementary combination in link-work the equivalent motion can be obtained by wheels in rolling contact.

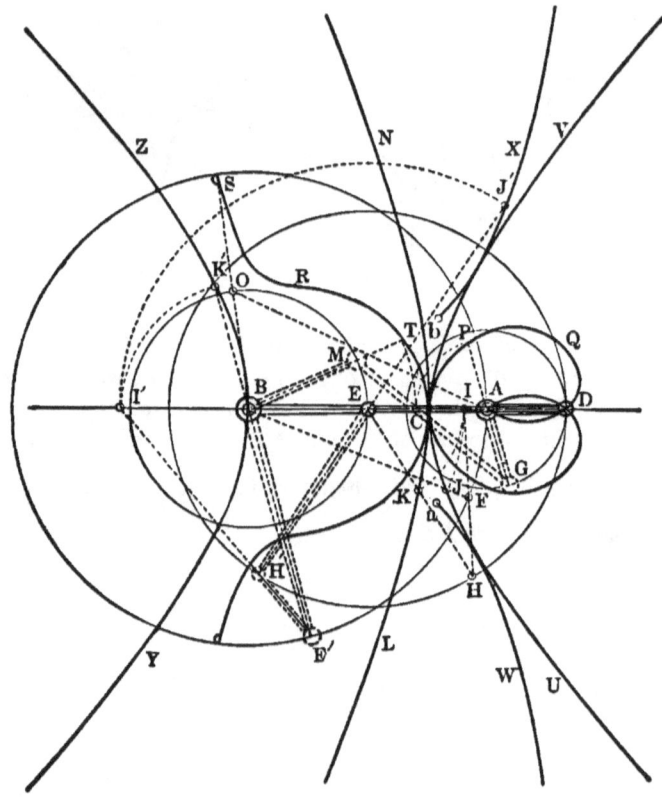

Fig. 297.

Example 1.—In Fig. 297 the axes A and B are connected with link-work as shown, AG being the driving crank, MG the connecting rod, and BM the driven crank. Their lengths are such that they will all come to coincide with the line of centers $DACEB$.

To find a rolling curve possessing the same law of velocity-ratio as the link-work, the links are to be put in several positions, as $AGMB$, etc., and the intersections with the line of centers found as at C for the position GM. Now the velocity-ratio in link-work being the inverse ratio of the segments AC and BC, also the same for rolling wheels in contact at C, it follows that a pair of wheels in rolling contact at C have the same velocity-ratio as the links $AGMB$, with intersection of link at C. Therefore, revolving C to T in BM prolonged, and C to P in AG prolonged toward P, we have points T and P in the peripheries of the rolling wheels. Other points, Q, D, A and R, S, are found in the same way, when with points sufficient the curves may be drawn in complete, above and below the line of centers, giving the rolling-wheel equivalents of the link-work.

The positions of these curves in the figure are proper to the crank positions of AD and BE, as if the wheels were finished and

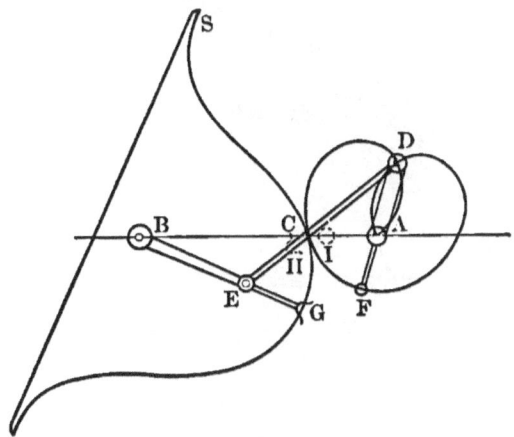

Fig. 298.

made fast to the cranks in these positions. The distance $BS = BA$, where the rolling suddenly stops, corresponding with the limit of the crank movement where the crank AD and link take positions on the straight line AO, BO being the extreme position of the crank BE when $AG + GM = AO$.

In Fig. 298 is shown as separated from Fig. 297 the link-work and the equivalent wheels due to the intersections of the link DE,

266   PRINCIPLES OF MECHANISM.

with the line of centers. Throughout the movement, the curves will remain continually tangent to each other at a point C, so that the point of tangency, and the intersection of DE with AB will remain some common point C on the line of centers.

In Fig. 298 the intersection C is for the line DE with the line of centers AB. If BE be fixed for the line of centers, the inter-

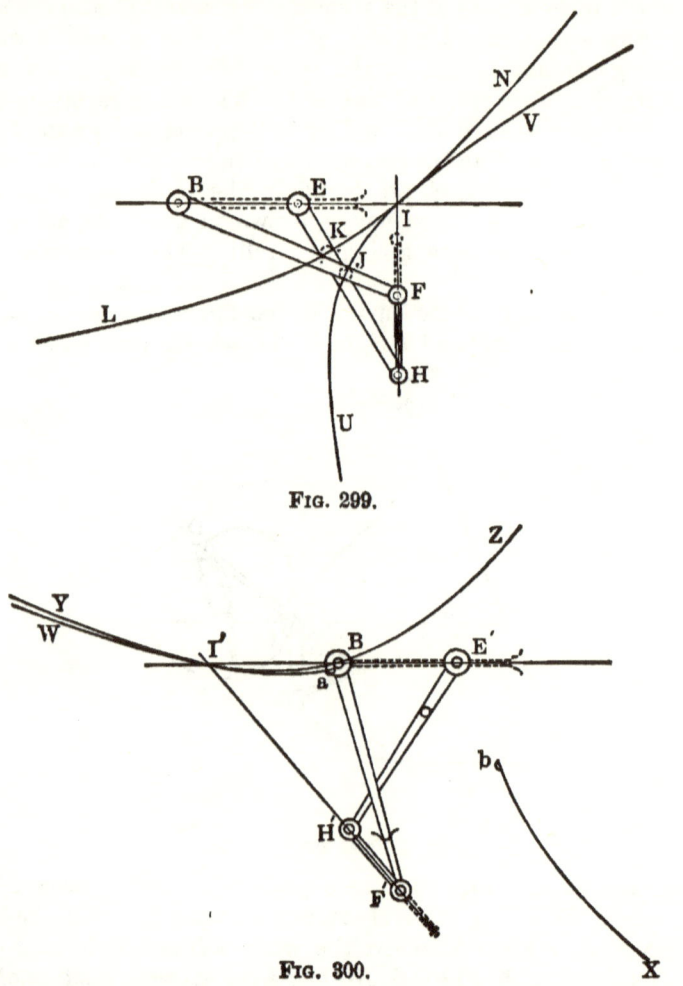

Fig. 299.

Fig. 300.

section will be that of the lines AD and BE, both prolonged, the curves for which are LKN, UJV, Fig. 299; and Wa, Xb, and YBZ, Fig. 300, as given in Fig. 297 and shown separately in Figs. 299

and 300, all extending to infinity, due to *FH* and *F"H'* becoming parallel to *BE*. One position of the links is shown at *BFHE*, Fig. 299, and another at *BF"H'E'*, Fig. 300, the first giving an intersection at *I* and the second at *I'*. The intersection *I* is revolved to the positions *BF'* and *EH*, Fig. 297, for points in the curve, while the point *I'* is revolved to *BF"* and *EH'* extended backwards, because this part of the curve is at the other side of the infinitely distant points. Thus Fig. 297 embraces the three sets of wheels of Figs. 298, 299, and 300.

Example 2.—The rolling-wheel equivalent of the crank and pitman is shown in Fig. 301.

The crosshead *A* as driver has its center of motion at an infinite distance, so that the line of centers is *BH* extended to infinity.

Taking the crank in the position *BG* and the pitman at *AG*, the intersection of the latter with the line of centers is at *D* in *AG*

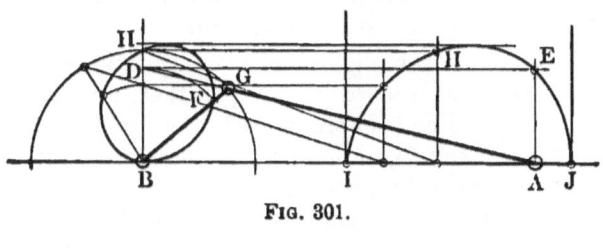

FIG. 301.

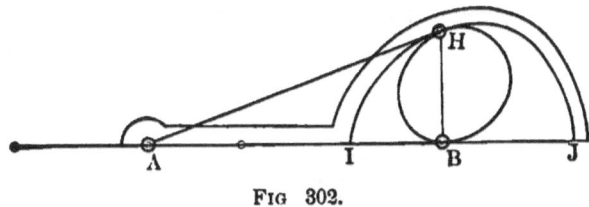

FIG 302.

prolonged. Revolving *D* to the crank gives *F* for one point in the wheel *B*. Also revolving *D* to the line *EA*, representing the crank line from infinity, gives the horizontal line *DE* and the point *E* in the mating wheel *A*. Proceeding thus, we obtain the curves *BFHB* and *JEHI* for the rolling-wheel equivalent for the crank and pitman motion for a half-turn of the crank. The proper position of the crank relative to the wheel *BFHB* is the line *BH*. The curve *IHJ* is supposed to be made fast to the crosshead and moving with it, as shown in Fig. 302.

Example 3.—In Fig. 303 we have an example of two cranks connected by a connecting rod and drag link, the connecting rod

Fig. 303.

having a fulcrum pin near its center carried on a swinging link, the lower end of which works on a fixed pin.

For this somewhat complicated elementary linkage the rolling wheels are worked out and teeth set upon them for a pair of gear wheels, the same being combined with the linkage in the one movement for a practical illustration of the equivalence of linkwork with its proper rolling wheels.

The gears show that the velocity-ratio of the linkage is far from constant, though, judged from the link movement alone, might be taken to be nearly so. By measured radii, taking the driver to be moving at a constant rate, the ratio of the fastest to the slowest for the driven wheel is over 3.

Example 4.—An interesting linkage is found in the Peaucelliers

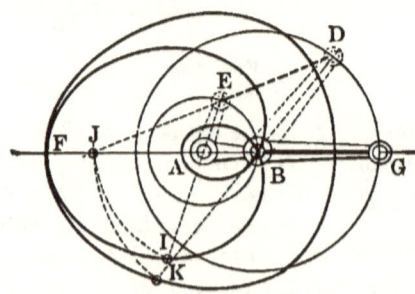

Fig. 304.

parallel motion, called by Prof. Sylvester the "kite," shown in Fig. 304 in the links *ABDE*.

Making AB the fixed line of centers, and A the driver, E describes the circle EB about A, and D describes the circle through DG about B.

Dotted positions of the links are shown to indicate the nature of the motion. The crank AE, it is readily seen, makes two revolutions to one of BD, and there are two positions for dead points, viz., when the joint D is at G, and at 180° from G.

The equivalent rolling wheels are readily found, one point for each wheel being at the intersection J of DE prolonged, J being its correct position for A, with reference to the crank position AE, and also the correct position for B relative to the crank position BD. The curves being symmetrical, we may revolve J about A to I and J about B to K in AE and BD prolonged, giving points I and K in the wheels relative to the cranks for the positions AB and BG, as if the wheels were cut in material and made fast to the cranks, the smaller one to crank AB, and the larger to the crank BG.

Thus determining a sufficiency of points and locating them all with respect to some one position of the cranks, we obtain the outline of the wheels shown.

The wheels fully worked out are shown together with the linkwork in Fig. 305, where the B wheel is seen to make one complete circuit of the rim with a side offset in it suitable for the mating wheel, and where the A wheel makes two complete convolutions, one being out of the plane of the other to prevent interference and mating respectively with the offset parts of B.

The wheels in this construction work by rolling contact of pitch lines, except at the dead points, where half-teeth are introduced as shown.

FIG. 305.

Example 5.—The case of two equal cranks revolving in opposite directions and a connecting link of the same length as the line of centers, but longer than the crank, is shown in Fig. 133, accompanied with the equivalent wheels, the latter being the rolling ellipses. Hence the law of velocity-ratio for this linkage is the same as that of a pair of rolling ellipses, each with the axis at a focus.

Example 6.—The same linkage as in Example 5, except that a shorter instead of a longer member is fixed to serve as a line of centers, is shown in Fig. 306, where the equivalent rolling curves are hyperbolas instead of ellipses. The hyperbolic wheel ECG is fixed upon the arm BE, while FCH is fixed on the arm AD.

For the positions AI and BJ of the arms these hyperbolas are tangent at the point K of intersection of IJ with the line of centers

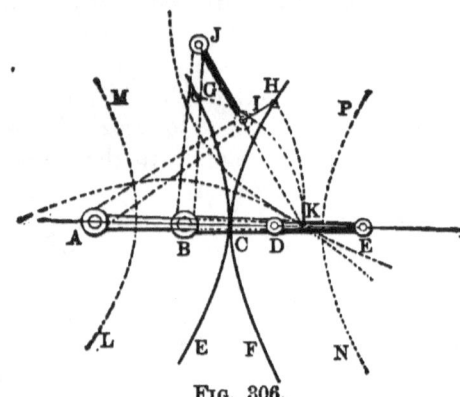

Fig. 306.

AB, both prolonged. The points H and G in the hyperbolas are obtained by revolving K to the radii AI and BG. These hyperbolas will roll in mutual contact from where the link IJ is parallel to the line of centers, through 180 degrees, to where it is again parallel to the line of centers, when another pair of hyperbolas, shown dotted, will come into rolling contact for the remaining 180 degrees.

The dotted hyperbolic half-wheel LM is to be fixed upon the arm AD, as well as the half-wheel FH, while the remaining two half-wheels, EG and NP, are to be fixed upon the arm BE, and at a distance between vertices equal AB.

The points B and D are foci to the hyperbolas EG and FH. Also A and E are focal points for the dotted hyperbolas. Thus A and D are focal points for the hyperbolas FH and LM, the asymptotes for which intersect each other, and the line AD at its middle point.

A pair of these hyperbolic wheels mounted for non-circular wheel pitch lines is shown in Fig. 49, where AB, as in Fig. 306, is the constant difference of lines drawn from all points C to the focal points.

The elliptic wheel of Example 5 and Fig. 133 may coexist along with the hyperbolic wheels of Fig. 306, except that one of the ellipses would be fixed with A and B as focal points, while the other would be carried on the link *IJ*, their point of tangency being constantly the intersecting point of *AI* with *BJ*, as shown in Fig. 307, while the hyperbolas are tangent where *AB* and *JI* intersect at *D*.

A most interesting and instructive figure in this connection is brought out by aid of his theory of *Centroids* by Reuleaux, and given in his *Kinematics of Machinery* by Kennedy, p. 194, much as shown in Fig. 307, where the linkage, the rolling ellipses, and hyperbolas are all presented in one view.

Thus with *AB* the line of centers, *IJ* may be a link connecting

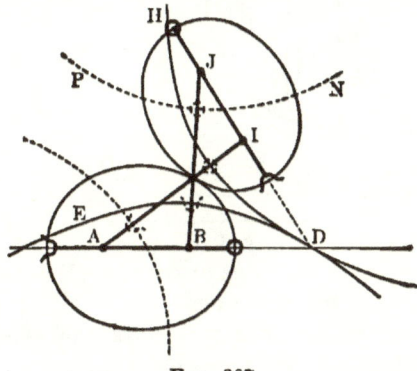

Fig. 307.

the focus *J* of *PN* with the focus *I* of *HD*, as has been admirably shown by Geo. B. Grant in his handbook on *Teeth of Gears*, Second Edition.

Examining Figs. 133 and 307, we find that for the ellipses both the link and line of centers in length equal the sum of the distances from the point of tangency of the rolling ellipses to the pair of foci *A* and *B* on the opposite sides of the center of the ellipse. In the hyperbolas the link or line of centers equals the difference of these distances from the point of tangency.

Referring these curves to their conic sections, we find that the parabola lies intermediate, and is like the ellipse of infinite length, or the hyperbola of infinite distance between foci.

Example 7.—Therefore we may expect that the link connecting foci of the parabolas would be of infinite length, as shown by Geo. B. Grant in his *Teeth of Gears*, Second Edition, and may be realized in the manner of Fig. 308, where *AB* is the fixed line of centers, *C* the point of contact, *A* and *D* foci, *DE* a link sliding on a straight guide, mounted on the parabola *A*, perpendicular to its geometric axis and giving to D the same motion as if DE extended to infinity and were pivoted there to the axis or opposite focus of A. The parabola D slides on a guide FG perpendicular to the

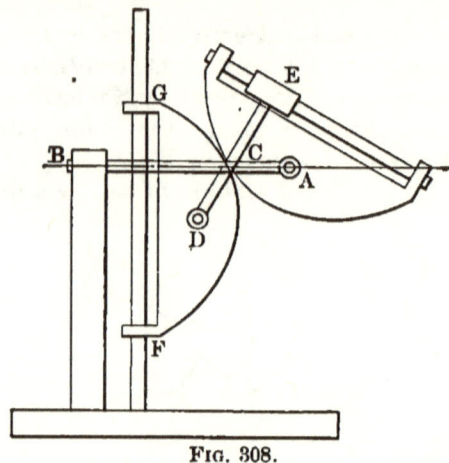

Fig. 308.

geometric axis of D. The point of contact C is at the intersecting point of the link and line of centers.

In Fig. 309 we realize this link-work movement without the parabolas. The gabs and pins will be explained later.

Example 8.—The above examples are all for curve equivalents

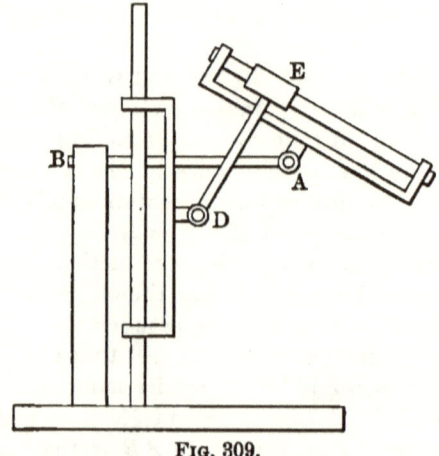

Fig. 309.

that are symmetrical; Example 2, apparently not, becomes so when the remaining half-revolution is provided for.

THE ROLLING CURVE EQUIVALENT FOR LINK-WORK. 273

As an example entirely wanting in symmetry, take the link-work of Fig. 293, the rolling-curve equivalents for which are shown in Fig. 310.

The links are shown in the position such as places the joint E

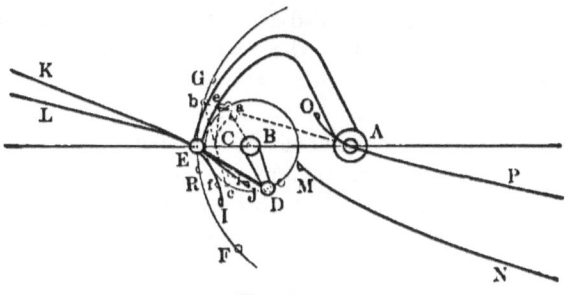

FIG. 310.

on the line of centers AB, and in the proper relation to the curves for mounting them upon the crank arms, IL and OP on AE, and JK and MN on BD.

To find a pair of points in the curves JK and IL: Place AE in the position Ab, when the link ED will follow to the position bc, giving the intersection C with the line of centers. Then, as before, revolve the point C to the arm Ab, giving the point e. Now if the curves were symmetrical this point could be retained as a point in proper location of the curve. But here e must be placed at f symmetrically on the opposite side of the line of centers, so that when the arm moves from E to b this point will move up to the line of centers. Also, the mating point a, is to be placed at i, a being on a line Ba, where the angle $CBa = DBc$. In like manner find other points.

These curves reach infinity when the link DE becomes parallel to the line of centers, following which the curves MN and OP come in from infinity. The point of OP at A is in contact with its mate when the link DE is in a line running through A; the portion AO answering to the slight return of D from its extreme point, as E moves from the straight line DA toward the extreme point F. For the other extreme, G, of movement of E, the points I and J are in contact on the line of centers. The movement of D is slight for the movement of E from R to F, while for RG the crank B makes over a 90-degree movement. The point R is about midway between F and G, so that D is comparatively quiet for half the time.

Dead Points in Link-Work.

A dead point or dead center, in link-work mechanism, is a point, or set of points, or positions of the links, at which, if certain of the links in combination be made driver, the linkage will be found positively locked. Thus when the crank and pitman are in line the crank cannot be started into motion by force applied to the crosshead. Dead points must always be provided against, either by inertia, by springs, or by extraneous attachments.

In the steam engine the inertia of the fly-wheel serves. In some reverse motions in machines springs are employed. In starting inertia is dead also; and so the single-acting steam engine must not stop on the dead center. In locomotives, two sets of cranks

Fig. 311.

are placed at right angles, so that one is at its best advantage when the other is on its dead center. Any angle will serve with a degree of efficiency, as illustrated in Fig. 311 of Boehm's movement, where an extra link is added to destroy the dead point. Several links may be added.

Another arrangement is shown in Fig. 312, where an extra crank

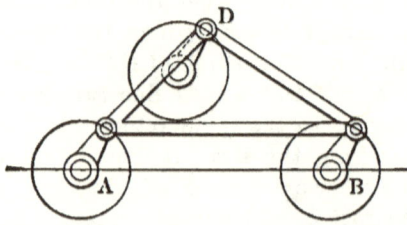

Fig. 312.

D, of radius equal that of A or B is added, together with the side extension AD and DB to the main link AB. The velocity-ratio is constant in the locomotive and in Figs. 311 and 312, the latter being obtained from Figs. 133 and 307 by cutting away the wheels and making the link parallel to the axis.

In the steam engine two pitman links at about right angles are sometimes connected on the same crank pin, thus realizing conditions equivalent to the use of two cranks with parallel rods.

Prof. Reuleaux has introduced a *gab* and *pin* at the points of the equivalent rolling curve where they are tangent to each other, when the linkage is at the dead centers. Thus in Fig. 306 a gab and pin is to be placed at C, where the hyperbolas are tangent, one on the link AD and the mate on the link BE.

In case a longer link is fixed for the line of centers the gab and pin attached to it become fixed also, as in Fig. 313.

Also, a gab and pin may be placed where the ellipses become tangent for the position of the line of centers, as shown in Fig. 314.

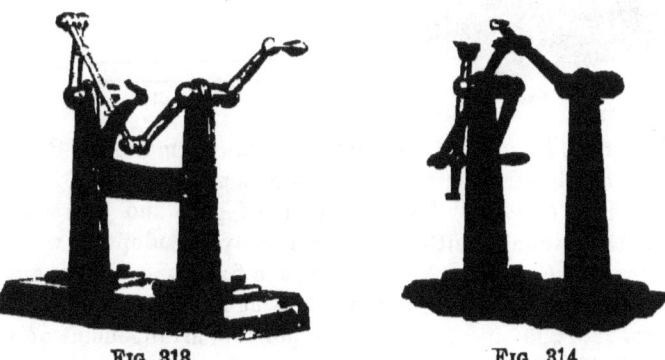

FIG. 313.   FIG. 314.

Here also a gab and pin become fixed when a shorter link is fixed for the line of centers, as shown in Fig. 315.

In Fig. 133 the gab and pin are in the form of a gear tooth and a space for the same, and represent the same case as Fig. 314, except that the elliptic wheel equivalents of the link-work are present and mounted with their respective links.

In Fig. 305 a half gear tooth and half-space are made to serve for gab and pin to carry over the dead center.

In Fig. 308 a gab and pin may be placed either at the vertices of the parabolas, or upon the links, as shown in Fig. 316, one set being sufficient. The dead center occurs when the swinging piece is vertical, since at that position the parts DE, FG, Fig. 308, may move up or down without control. One set of gabs and pins between A and D will prevent this.

In Fig. 297 the curves show that a gab and pin may be placed at C as indicated by either system of curves. In Fig. 298 the pin, for instance, may be placed on AD extended downward from A to the curve at F, while the gab may be placed on the line of BE extended to the curve at G. Thus the gab and pin are placed on the

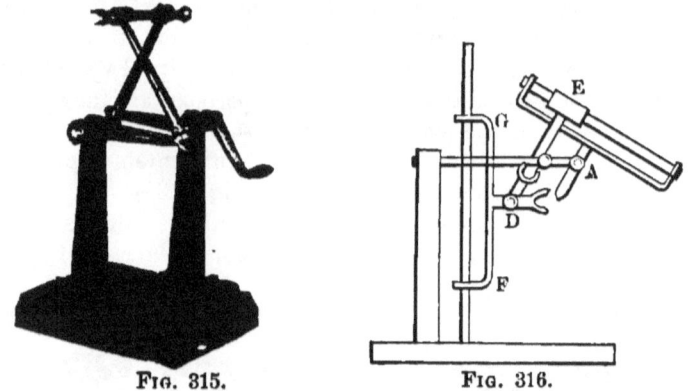

FIG. 315.   FIG. 316.

crank arms AD and BE, so that both are in motion. But a pair may also be placed on the longer members as at H and I, where the latter is fixed on AB. Thus two pairs of gabs and pins are available in this linkage, either of which may be adopted, whichever member, as AB, AD, DE, or BE, is the fixed one, and in any case one, a gab or pin, may become fixed, as at I.

Figs. 298, 299, and 300 show the possible arrangement of all the gabs and pins for this linkage.

The linkage of Fig. 307 shows eight gabs and pins, any two pairs of which may be adopted in a particular case—sometimes all in motion, and sometimes two being fixed, as illustrated in Figs. 313 to 315.

In Fig. 316 four of the gabs and pins are at infinity; while of the four shown, one is fixed.

Path of the Gab and Pin.

In Fig. 317 is shown a link and crank connection recently adopted with success into a machine to transfer motion from one shaft to another parallel to it, the gabs and pins being employed. The connecting link is raised from its position a distance $FG = DE$ to uncover the gabs at a and b.

The curves described by the centers of the pins a and b are drawn in, to show how the gabs must widen in amount of opening. The pins *a* and *b* are placed one-sided, for the reason that, if placed central, they would interfere with the gabs. But an examination of the path curves for *a*, *b*, and *c* shows that these curves are cusp-

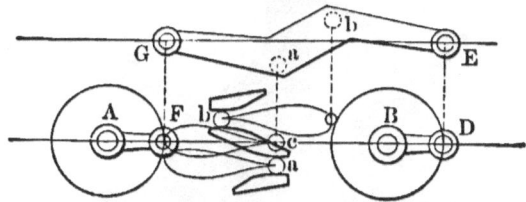

Fig. 317.

shaped at the gab positions, and that the central one, *c*, does not differ materially from those at the sides at *a* and *b*.

As to the advantage or disadvantage of placing the gabs somewhat off of the line of centers, very little difference will be noticed.

The position *c* is right for the rolling ellipses of Fig. 133, but here the eccentricity is so slight that interference occurs.

The path curves for *a*, *b*, and *c* were traced by means of a paper templet on the drawing board, the templet being so cut that the edge fitted at the points *F*, *a* and *D*, and had a mark at each. This templet, placed in the various positions of *F* and *D* in the circles, and *a* noted for each, gave points through which the curves were traced.

An example of a linkage nearly like that of Fig. 299 in application, is given in Fig. 318, which picks up the staple at the lower

Fig. 318.

end of the machine and delivers it at the upper end, thus representing the handling of rods, screws, etc., in the manufacture of those articles.

278   PRINCIPLES OF MECHANISM.

In Fig. 319 is given a model of link-work serving to prove that vibratory motion may be multiplied thereby. Thus the pointer is given four movements, to one of the first vibrating bar of the series.

Fig. 319.

CHAPTER XXV.

LINK-WORK—(Continued).

II. Axes Meeting.

THIS is sometimes called conic link-work, or solid link-work, the principal essential consisting in bringing all the axial lines of shafts

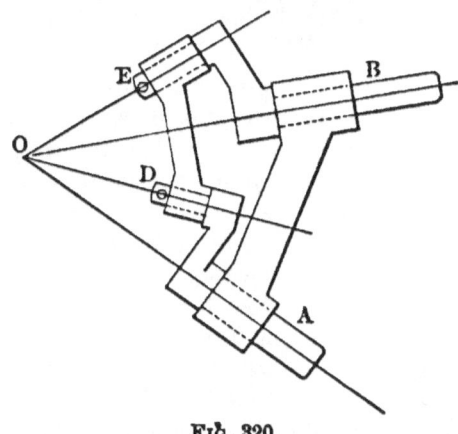

FIG. 320.

and pins to a common point, O, as in Fig. 320. Thus we may have equal cranks, as in the counterpart for parallel axes of Figs. 306 and 307; or unequal cranks, as in Fig. 299.

To realize the principles of equal cranks, it is only necessary to gives the cones AOD and BOE an equal slant.

Any of the examples under axes parallel may be carried into conic link-work, even to the extent of continued trains.

The Velocity-Ratio in Conic Link-Work.

It will be advisable, if not necessary, to refer problems in conic link-work to spherical surfaces normal to all the axes, or which have O, Fig. 320, for the center of the sphere.

In practice, avoiding spherical trigonometry, the velocity-ratio

280   PRINCIPLES OF MECHANISM.

may be most readily determined by preparing the proper spherical surface in wood or other material, and using it for the drawing board. Take Fig. 321 to represent this drawing board and drawing.

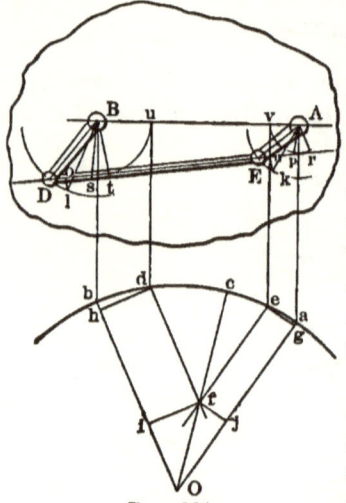

FIG. 321.

A and B are the points where the axes A and B pierce this spherical surface; D and E where the crank-pin axes pierce the sphere; AB is the line of centers and DE the line of the link, both being parts of great circles, and constant in length on the sphere.

Suppose A makes a slight turn, moving E to k, then D will move to l so far that the projections of Dl and Ek upon DE will equal each other and give Do equal to En, equal to pr, equal to st; since the spherical connecting link DE remains of constant length, s and p being points noted thereon where Ap and Bs are perpendicular to DE. Then the angle sBt measures the angular displacement of axis B, corresponding with that of pAr for A. But sB is a circle arc on the sphere, also Ap. Revolving s to u and p to v and transferring the arc Bu and Av to bd and ae in the great-circle section abO, we readily obtain the perpendiculars eg and hd, the inverse ratio of which is the velocity-ratio, because

$$hd \times \text{angle } sBt = st = pr = eg \times \text{angle } pAr,$$

giving

$$\text{velocity-ratio} = \frac{\text{angle } sBt}{\text{angle } pAr} = \frac{eg}{hd}.$$

Hence in practice, for any position of links as $AEDB$, draw the perpendiculars to the links Ap and Bs, which arcs transfer to ae and bd. Then the velocity-ratio equals the inverse ratio of eg and dh.

The Rolling-Wheel Equivalent of Conic Link-Work.

In Fig. 321 draw parallels ef and df to the axes Oa and Ob, and through f draw Oc. Then ac and bc will be the spherical arc radii for the spherical rolling-wheel equivalents of the link-work of Fig. 321, for contact occurring when the links are in the positions shown. When the links are located for drawing the wheel, as for instance AE on the line of centers, their radii may be laid off in place. Likewise for other radii, until a sufficient number of points are determined for drawing in the wheel.

The Dead Center and Gab and Pins in Conic Link-work.

In Fig. 322 is given a photo-process copy of an example of conic link-work for equal cone slants, and where the angle between the axes A and B is equal to that between the pins D and E when in one plane as in Fig. 320.

Fig. 322.

Thus conditioned, the cranks will turn continuously in the same direction as in Fig. 312, or in opposite directions, as in Fig. 313.

The dead center accompanies conic link-work, and gabs and pins may be used here as well as in the case of axes parallel. For Fig. 322, the elliptic conic pitch lines of Prof. MacCord will serve to determine the locations of the gabs and pins, those curves serving as the rolling-wheel equivalents of this linkage.

Other linkages may have the rolling-curve equivalents determined as in Fig. 321, when the location of the gabs and pins is readily made.

EXAMPLES OF PECULIAR MOVEMENTS.

Example 1.—In Fig. 323 is what we might term a bent-shaft movement. A plunger is made to slide in the top head in a direction parallel to the shaft.

The joint work could be simplified by using a block and two pins, as in Fig. 324. If B is a square bar the center line of the pin D should strike the intersection O.

The velocity-ratio is the same as for the swash-plate movement shown in Willis, p. 172, where it is proved that the motion of the bar is the same as that of the crank and crosshead, with infinite pitman.

The motion is still the same, if in place of the angular part of the shaft in Fig. 323 an enlarged bearing were used, like an ordinary eccentric and strap, and mounted on the straight shaft central, though at an angle like that of the swash plate.

A movement like this has been used for working the valve in a steam engine exhibited at the Centennial of 1876.

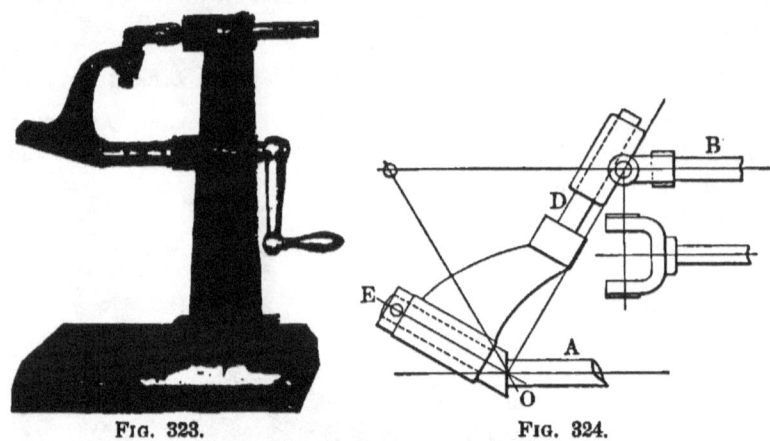

FIG. 323.   FIG. 324.

Example 2.—The Hooke's universal joint is often employed as a shaft angle-coupling, where the velocity-ratio is an important consideration.

The joint is shown in Fig. 325, where A and B are the shafts to be connected, AD and FBE half-hoops between which is a cross with one branch at EF, parallel to the paper, and the other at D perpendicular to the paper. The branches of the cross are pivoted at E and F, and at the two points D, in AD.

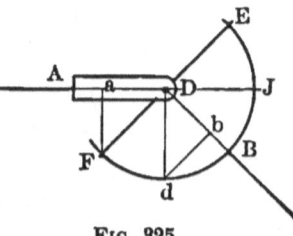

FIG. 325.

In the position shown the velocity of A is greatest, and the

$$\text{velocity-ratio} = \frac{DF}{aF},$$

while at a 90-degree turn from the position shown the velocity of B is greatest when the

$$\text{velocity-ratio} = \frac{db}{Dd},$$

II. AXES MEETING. 283

because for the first position A acts in effect by an arm aF upon an arm DF from B, since aF is continually in the plane of the axis of A, and of the arm DF of the cross.

If A revolves uniformly, the ratio of the fastest for B to the slowest, since $db = aF$, is

$$\frac{\text{Fastest for } B}{\text{Slowest for } B} = \left(\frac{DF}{aF}\right)^2.$$

These limiting speeds are in the same ratio as those for elliptic wheels of Fig. 133, where DF and AF are the distances from a focus to the remote and a nearer vertex respectively, though in the latter we have but one max. and one min. speed in a revolution, while in Fig. 325 we have two. But the law of variation of velocity between extremes is not the same as for the elliptic wheels.

In the case of a pair of two-lobed elliptic wheels, with the maximum and minimum diameters in the same ratio as DF to aF, the ratio of the limiting speeds is the same as for Fig. 325, and the number of changes in a revolution is the same, but the law of variation of velocity still remains different.

A pair of overhead shafts A and B connected by this joint, when the angle ADB differs much from 180 degrees, will be accompanied by too great a variation of velocity-ratio.

In this case the use of two joints between A and B, arranged as in Fig. 326, with the branches EF and HI of the crosses in the same plane and with the angle $AOB = 2\ GDO$, will serve to transmit motion from A to B, with the velocity-ratio constant.

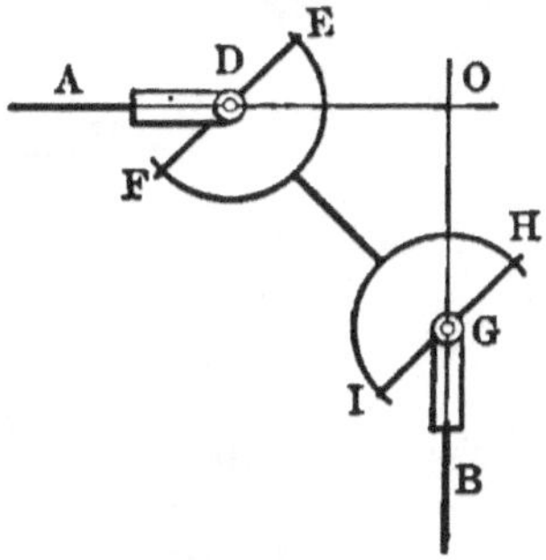

For the single joint of Fig. 325 the law of velocity-ratio is given by the formula brought out in connection with Fig. 256, viz.,

FIG. 326.

$$\frac{\text{ang. velocity of } A}{\text{ang. velocity of } B} = \frac{1 - \sin^2 x \cdot \sin^2 BDJ}{\cos BDJ},$$

in which x is the angle of movement of A from the positions shown.

This equation may be constructed and solved graphically by the diagram, Fig. 327. Thus, draw BDJ with the same angle, BDJ, as in Fig. 325. Lay off $DJ = 1$ by some scale, and draw JL and LM perpendicular to DL and DM respectively.

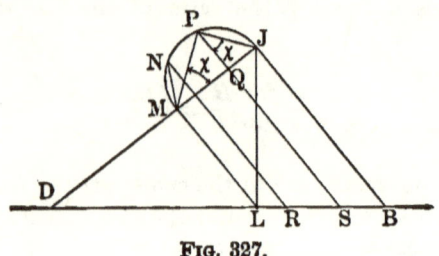

Fig. 327.

Then
$$JL = \sin BDJ$$

and
$$JM = JL \sin BDJ = \sin^2 BDJ.$$

Draw the semicircle JPM and lay off any angle $x = JMP$. Then
$$JM \sin x = JP$$

and
$$JP \sin x = JQ = JM \sin^2 x.$$

Then
$$DQ = DJ - JQ = 1 - JM \sin^2 x = 1 - \sin^2 x \cdot \sin^2 BDJ.$$

Draw the line PS, and we have
$$DS = DQ \sec BDJ = \frac{DQ}{\cos BCJ}.$$

Hence the

velocity-ratio of A to $B = DS = \dfrac{1 - \sin^2 x \cdot \sin^2 BDJ}{\cos BDJ}$.

This supposes $DJ = 1$, but if it equals any other value, measure DJ and also DS by the same scale, and divide DS by DJ for the velocity-ratio.

In a practical case draw BDJ, JL, LM and the semicircle JPM, as explained. This much is constant for a particular case of an angle BDJ. Then lay off all angles x the velocity-ratio is desired for, drawing the lines MN, MP, etc. Then project lines NR, PS, etc., perpendicular to DJ when we have the series of velocity-ratios DR, DS, etc.

The velocity-ratio DL is a minimum and BD a maximum; the latter answering to the position shown, where $x = 0$.

If A revolves uniformly, the fastest for B divided by the slowest gives

$$\frac{\text{Fastest for } B}{\text{Slowest for } B} = \frac{DB}{DL} = \left(\frac{DJ}{DL}\right)^2 = \left(\frac{DF}{aF}\right)^2, \text{ (Fig. 325)}$$

since from Fig. 327

$$DL : DJ :: DJ : DB,$$

or

$$DB = \frac{\overline{DJ}^2}{DL} \text{ and } \frac{DB}{DL} = \left(\frac{DJ}{DL}\right)^2.$$

Thus the results of Fig. 327 agree with that obtained from Fig. 325.

This example, the Hooke's joint, is found in various forms of construction, the simplest being that for couplings for *tumbling rods* for transmitting power from the "horse-power" to the threshing-machine in agricultural districts. On each end of each rod is a forked casting, much like that in Fig. 324, with bosses through which a pin may be placed at right angles to the rod. A block goes loosely between, with holes at right angles and just missing each other. Two pins or bolts are used at each joint, each passed through a fork and the block, and at right angles. Thus, between adjacent rods is a universal joint, so that the series of rods may lie upon the ground upon notched blocks, or on blocks and between stakes.

For a considerable angle between rods at a joint some end play will occur, when the holes for the bolts, as above, pass beside each other.

Example 3.—The Almond, Reuleaux, and other joints or couplings, are more compact than the Hooke's, so that they may very readily be enclosed in an oil-tight case, to facilitate lubrication.

In Fig. 328 is the Almond coupling, serving to couple a pair of axes A and B at an angle, and with constant velocity-ratio equal to 1.

At D is a fixed shaft, at the intersection of, and perpendicular to the plane of the axes A and B. On this shaft a sleeve slides

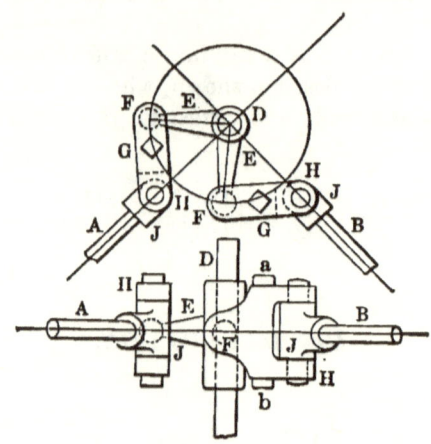

Fig. 328.

from which project two arms E, E, subtending the angle ADB, and upon the ends of each of which arms is a ball to work in a socket F in the swinging piece FH, the latter being in two pieces, held together as one by a bolt ab, and pivoted to J.

When A revolves, each ball F is compelled to travel in a curve on the cylinder whose axis is D, one curve being identical with the other. Since both joined pieces FH are identical, it follows that the motion of B is a copy of that of A, and hence the velocity-ratio equals 1.

This joint or coupling will connect shafts at any angle ADB, provided the angle between the arms EE is made the same. If the shafts A and B are in one straight line, the coupling will reverse the directions of motion.

In Fig. 329 is a somewhat different joint or coupling, where the shafts A and B have fixed cranks with long crank pins, extending through the sockets or sleeves F, the latter having right-angled pin holes passing without meeting, through which are pins fixed in the arms E of the central piece. The latter slides on the fixed shaft or stud D, placed as before at right angles to A and B, and at their point of intersection.

It is readily seen that the parts D, EE, and FF compel the crank pin B to maintain the same relation to the shaft B as the crank pin of A does to the shaft A, so that the velocity-ratio is constant.

The shafts may here also be connected at any angle from 0 to 180 degrees, or until interference of cranks occurs, the angle between the branches EE being made the same as that between the shafts A and B.

Besides this, the shaft B may be placed at any height above or below A by simply making D of suitable length and placing the arms E connecting with B the same amount above or below those

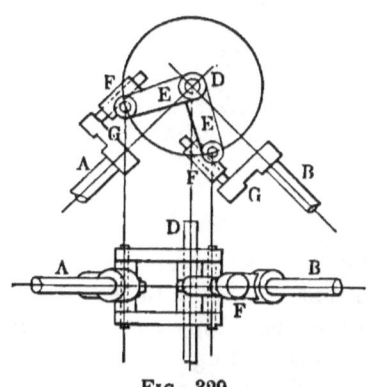

Fig. 329.

connecting with A as the one shaft is above or below the other. Thus by making D open, and extended into sliding and revolving gudgeons above and below, we meet the case of connecting by linkwork axes which cross without meeting, and with a velocity-ratio unity.

The Reuleaux coupling is shown in Fig. 330, where A and B are the shafts connected, C a head on their ends with holes at right angles to A or B for a joint pin, the latter passing the forked ends of EF at F, permitting the ends E to swing freely, which ends E are sleeve-like, receiving the round prongs of the head D, thus connecting EE through D. The ends of the prongs of D nearest F may have a nut and collar to prevent D from being thrown outward in revolving.

This coupling or joint, so put up as to have the angle AFF equal the angle FFB, will have a velocity-ratio constant.

This, as well as Figs. 327 and 328, should have firm bearings for A and B, close to the movement, also in the former ones for D, an advisable arrangement being an iron framework especially for them.

One important practical advantage of this latter over the former

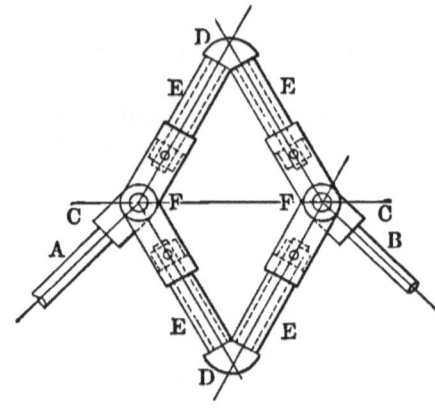

Fig. 330.

two couplings is the fact that the mutual sliding of parts is simply from rotation upon pins, according to the true ideal in link-work, while in the former we have this combined with a disproportionately large amount of end sliding.

At F the branches of each part EF may be made unequal and one piece EF like another, so that the bosses of one piece EF will lie beside those of the other, uniting at F; and still admitting the four sockets E to all lie in one plane $EEEE$.

In Fig. 329, as suggested by Prof. E. A. Hitchcock, we have a means of connecting the shafting in one shop room above or below another by link-work.

In fact we may say any number of shafts parallel to one another in any one plane, by means of one shaft D, Fig. 329, parallel to the plane, crossing all the shafts at any angle.

This last statement corresponds with placing the shafts A and B, Fig. 329, with others in parallel in any plane, but at a distance from one another along the line of D, where the latter consists of a shaft free to slide endwise in bearings with a single-crank arm-piece E at each shaft A, B, etc., all keyed upon the shaft D in parallel, D being near the plane of the shafts and parallel to it. See Fig. 333.

CHAPTER XXVI.

LINK-WORK—(Continued).

III. Axes Crossing Without Meeting.

Ratchet and Click Movements.

In Fig. 331 we have the typical case of projection of axes which are not parallel and not meeting, where A and B, A' and B' are the axes, Da and Eb the crank or lever arms, and ab the connecting link. At a and b some sort of universal joints are necessary, as, for instance, the ball-and-socket.

According to Fig. 267 it would appear that the velocity-ratio may be obtained by giving the parts a small rotation, so that the

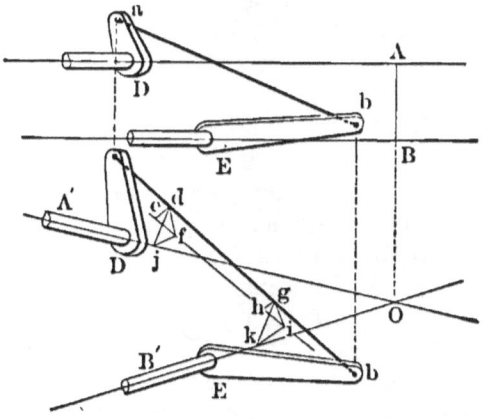

Fig. 331.

end d of the common perpendicular dj makes the movement df, which, projected upon the connecting rod, gives ef for its endlong component of displacement; also, that the end of the common perpendicular gk makes the movement gi, which projected upon the connecting rod gives hi for the endwise component of displacement, which components ef and hi must be equal, allowing permanence of length of connecting rod, so that the velocity-ratio will equal the ratio of the angles *djf* to *gki*, each determined as the angle between two meridian planes containing the radii *jd* and *jf*, or *kg* and *ki*.

To avoid the universal joint at *a* and *b*, Prof. Willis introduces a second intermediate piece between *a* and *b*, as at Fig. 332.

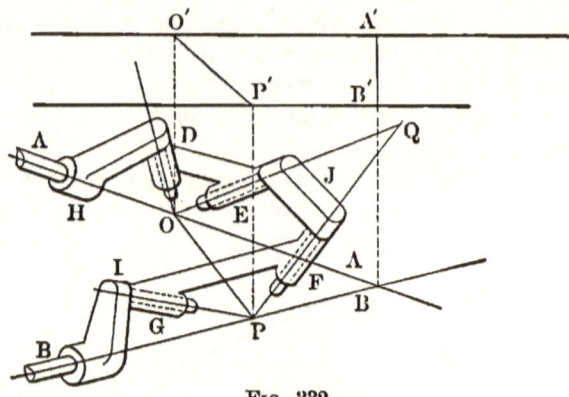

Fig. 332.

The shafts are shown as not parallel and not meeting, by the lines $O'A'$, $P'B'$ in elevation, and in plan by the lines OA and PB.

At H is a crank for A, and DE a connecting rod with the three axes meeting at O. At I is a crank and GF a connecting-rod, with the three axes meeting at P. At J is an intermediate piece, with axes QP and QO to connect DE and GF.

The parts thus connected may be moved about on their straight axes when O remains a fixed point of the intersection of the axes AO, DO, and QO, while P remains a fixed point for the axes BP, GP, and QP. Also, Q is the point of intersection of the axes of J. In use these parts have long bearing surfaces without cramping or endlong sliding.

This mechanism becomes somewhat complicated for connecting axes that do not meet. In many practical cases some sliding is not objected to, since the sliding facilitates lubrication, and by admitting sliding and rotary motion combined we often greatly simplify the mechanism, as in the following:

Examples.

Example 1.—In Fig. 333 we have the case of axes crossing but not meeting, and yet have but one piece intermediate between the cranks upon the axes A and B, that piece serving as a link of infinite length.

The crank pins are parallel to the axes which support them.

III. AXES CROSSING WITHOUT MEETING. 291

As the axes are both fixed in direction, it follows that the angle

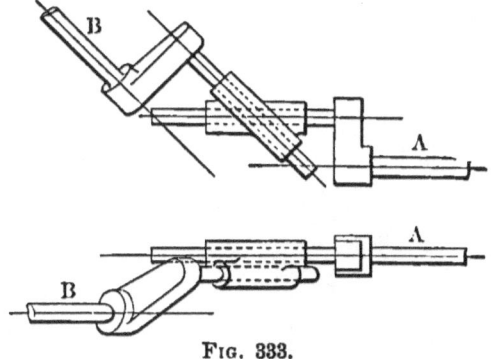

Fig. 333.

between the pins is fixed, and that a block with long bearing holes at the same angle may be employed to connect the crank pins.

Continuous motion is possible, except for the dead centers, and these may be passed by use of gabs and pins at the points determined in position, as by the equivalent non-circular gear pitch lines.

Example 2.—In Fig. 334 is given an example which has been adopted in certain machines, with hundreds of them in successful operation for some years past. The axes A and B are at right angles and not meeting.

The double sleeve D has holes at right angles fitting the pins closely. It turns out to be a simple, compact, and efficient movement, the piece D serving as a short link.

Example 3.—In Fig. 335 is an example, in practical use in machines, of a case of axes A and B at right angles and not meeting, the piece D working on an eccentric A as driver, upon which it slides to accommodate it to the arc F about the fixed axis B, while D slides upon E to admit of the lateral movement of the eccentric.

In practice, the longitudinal sliding of the knuckle piece D, combined with the rotary, is found an advantage, as favoring distribution of lubricant, and smoothness of surfaces in wear.

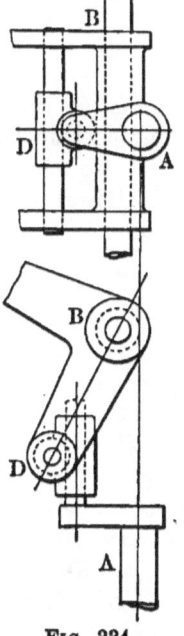

Fig. 334.

Example 4.—Here link-work, much like that of Fig. 295, drives a pin F' in a curve much like that of FE, while a lever pivoted at B carries a pin G, between which and F' is a knuckle piece D, working freely on both pins.

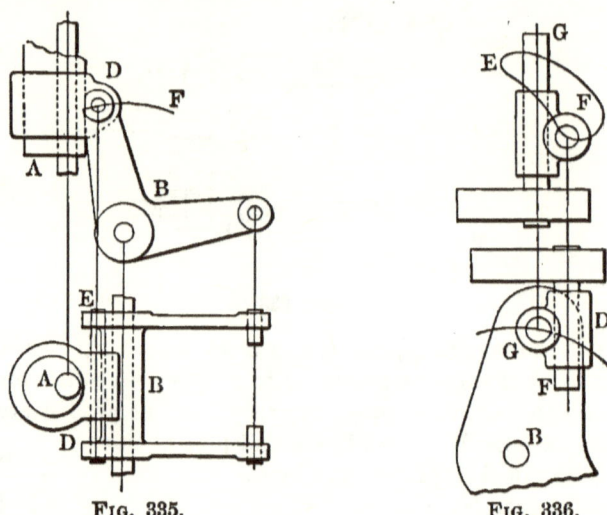

Fig. 335.   Fig. 336.

As the pins are always at right angles, the piece D may have long bearing holes at right angles.

If we take A of Fig. 295 as driver, and B, Fig. 336, as fulcrum of follower, these axes A and B cross without meeting. This is in successful operation in hundreds of tacking machines.

Ratchet and Click Movements, Axes Variously Related.

These movements are properly classed with link-work, sometimes called intermittent link-work, because one end of the click or pawl is usually supported on a pin, while the other in action rests in a notch, which serves nearly as if pinned at that point.

Example 1.—**A Running Ratchet** of simple form is shown in Fig. 337, such as was employed in the old-fashioned sawmills to feed along the carriage and log. A is a fixed pin about which AD swings, causing the end E of the click DE to move forward and back. As E moves forward the teeth of the wheel B are engaged and moved, while on the return E draws back over the teeth, the detent click F engaging a tooth and preventing the return of the wheel.

Arranged as in the figure, D moves in the arc of a circle which, together with the movement of EB, causes ED to change position relative to EB, giving greater liability of E to slip out for one position than for others. Also, this change of position, or swinging of E in the notch, gives cause for wear. To reduce this, the pin D should be so placed as to swing in a normal to a line AB, or better,

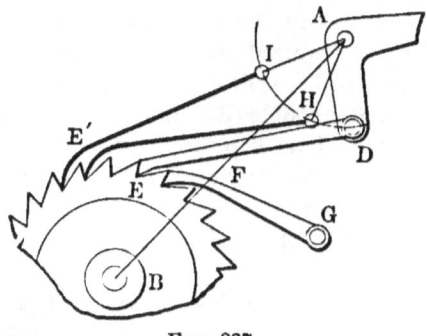

FIG. 337.

a circle about B. An approximation to this, while using AD, is obtained by locating D to swing from H to I, for which the position of DE with respect to EB is nearly constant for forward movement of E.

In other cases D is mounted on a pin centered at B, or on the axis B, so that ED is fixed relative to EB when in action, as in the case of Fig. 339.

Example 2.—**A Running Ratchet for Varied Step Movement** is shown in Fig. 338, as used in thermometer-plate graduating machines over thirty years ago. At B is a vertical axis about which the slotted table GH swings. The slots ab, cd, etc., each have a stud that can be made fast at any point in the slot. These studs are slotted to receive a flexible ratchet strip DF, and are made fast in studs by thumb screws, as shown at J. A click, EE, as long as the slots works the ratchet strip, and moves the table HG, notch by notch, between each of which move-

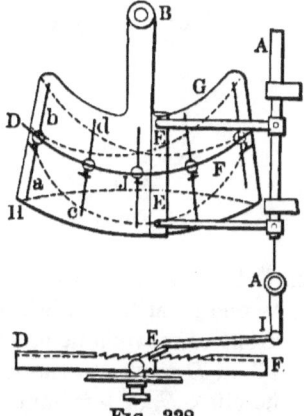

FIG. 338.

ments the graduating tool cuts a mark in the scale being graduated, the slide supporting the scale, tool, etc., not being shown, the scale and its slide being moved in a straight line proportional to the angular displacement of HG.

The ratchet strip DF is shown concentric with B, when it gives a uniform scale; but it may be set in any position, as dotted, on the table HG, when a correspondingly varied scale results—as finer at one end, or the other, or in the middle, or at both ends.

This variability is required to meet the case of glass thermometer tubes which cannot be made of uniform calibre.

Example 3.—**A Reversible Running Ratchet** is shown in Fig. 339, where the click A is kept in engagement either way by a spring action at C. A central notch for C holds the click out of engagement when desired.

A variety of forms of A are in use, especially in feed motions for machinists' tools.

Example 4.—**A Reversible Varied Rate Running Ratchet** is shown in Fig. 340, where G is continually reciprocating about B, carrying the pawls or clicks E and F forth and back on the click

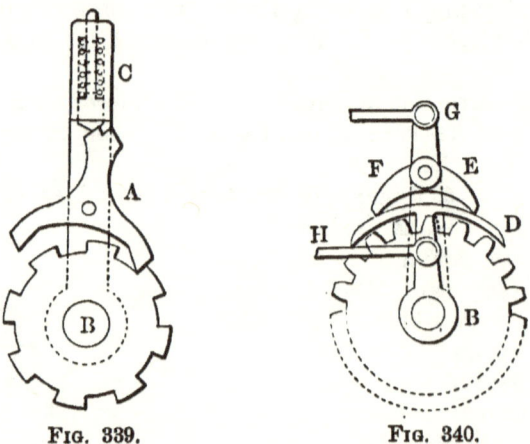

Fig. 339.   Fig. 340.

guard D. As H moves D to the one side or the other a click engages one or several teeth, and moves B little or much, according to extent of movement of D.

When D moves to the left the wheel B is turned right-handed by the click E, or left-handed by the click F if D is thrown to the right. The more D is displaced, the more rapidly will B be moved by the click, and when central, B is stationary.

This movement is found in Snow's Waterwheel Governor.

The click guard D has had frequent application with good results.

Example 5.—**A Continuous Running Ratchet** is shown in Fig. 341, where the wheel is moved the same way for each movement of the handle J one way or the other. Thus the holding or detaining click of Fig. 337 is here made a working click. The clicks may push or pull in action, the former being usually preferred for considerations of strength and shown above in full lines, while the latter are dotted in. The pins D and E are sometimes placed in a line perpendicular to AB, but the figure gives that relation which causes least turning of the ends of the clicks in the teeth, and consequently least wear.

The teeth may be internal as a possibility, but are usually external.

Forms of Teeth.

At D, Fig. 342, the normal d to the working surface a of a tooth

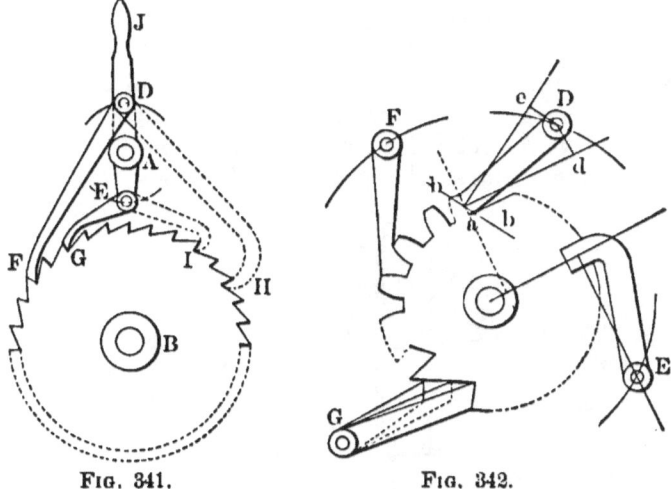

Fig. 341.   Fig. 342.

may go inside of D, when the click will be secure. But if it goes outside of D, as for the case that b were the working face of the tooth, the click will be insecure, or very likely to fly out of engagement.

Again, if Dd is excessive, the wheel will turn back appreciably before coming to bearing against the click after it drops in.

The click may have a shape for engagement with a common gear tooth, as at F.

At G several clicks may be placed on the same pin, and of dif-

296   PRINCIPLES OF MECHANISM.

ferent lengths, one engaging after another and serving in effect to divide the tooth pitch into parts answering to a finer toothed ratchet wheel. This may be called a *differential Ratchet*.

At E is a *stationary ratchet* click serving to hold the wheel stationary against moving either way.

Friction Ratchets.

In Fig. 343, E may represent a round rod and D a washer easily sliding over E, except when cramped, as by the pull F, when the

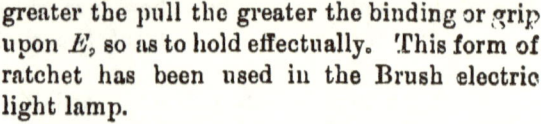

greater the pull the greater the binding or grip upon E, so as to hold effectually. This form of ratchet has been used in the Brush electric light lamp.

The part E may be a square rod or a flange on a wide piece, D fitting properly.

This form of ratchet always works satisfactorily when new, but in practice very soon becomes untrustworthy, and is usually abandoned, especially under heavy service.

FIG. 343.

Shoe pieces will extend the wearing surfaces, as in Fig. 344, and they have been used with fluted seats, as shown in plan.

These work admirably for a much longer time than the more limited bearing surfaces of Figs. 342 or 343, but finally become

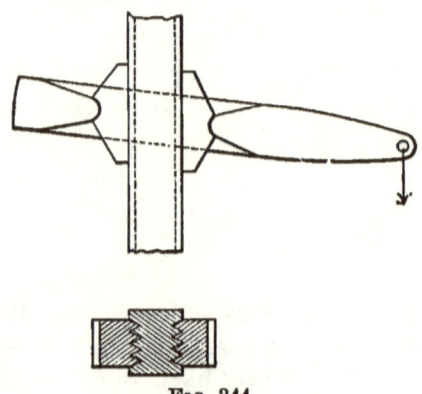

FIG. 344.

worn in places where most used, and fail to work as expected.

Example 1.—**A Continuous Friction Ratchet** is shown in Fig. 345, which was once used on a rock-drilling machine for feeding the working parts along as the drilling progressed. At D is a flange fast to the stationary framework, while *J* is a part of the sliding carriage, *G* and *F* being the grip clicks, one holding in the

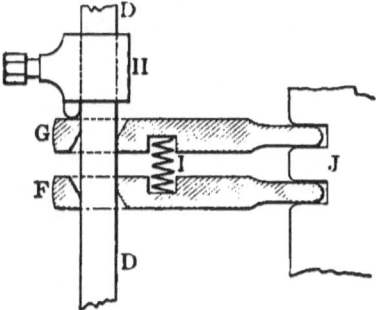

Fig. 345.

opposite sense of the other. *H* is a clamp under slight friction grip, that can be moved towards *G* as the drilling advances and a tappet comes to press upon it. That moves *G*, and in turn *F* moves by the spring *I*. *F* prevents *J* from moving up, and *G* opposes its downward movement until *H* acts upon it.

This apparatus operated most admirably for a few days, but soon had to be abandoned, as most ratchets of this kind.

Example 2.—**Wire Feed Ratchets**, as in Fig. 346, with teeth to grip the wire, the lower ones reciprocating together while the upper ones are stationary, always operate with satisfaction for a few hours, for feeding wire, when the teeth become worn and slip, making the contrivance a failure.

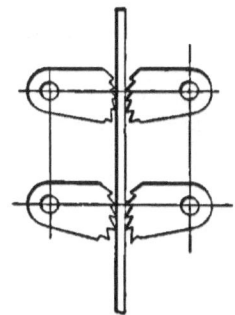

Fig. 346.

Example 3.—**Running Face Ratchets** have the teeth upon the sides of the wheel instead of edge, as shown in Fig. 347. In this example, recently adopted with success in practice, the click *E* is a full circle, as well as the wheel *D*, both having the same number of teeth, viz., 100.

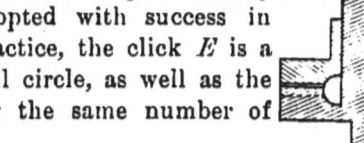

Fig. 347.

Ratchet Gearing is made in great variety—too much so to attempt full description here. An elaborate enumeration and illustration will be found in Reuleaux's *Constructor*, translated by Suplee.

PART V.

CHAPTER XXVII.

REDUPLICATION.

THIS term is applied to such movements as consist of circular pulleys or sheaves, and flexible connectors passing over them, usually made fast at one end, mostly ropes, employed to multiply the pull, or the motion. Blocks and tackles of ships are included.

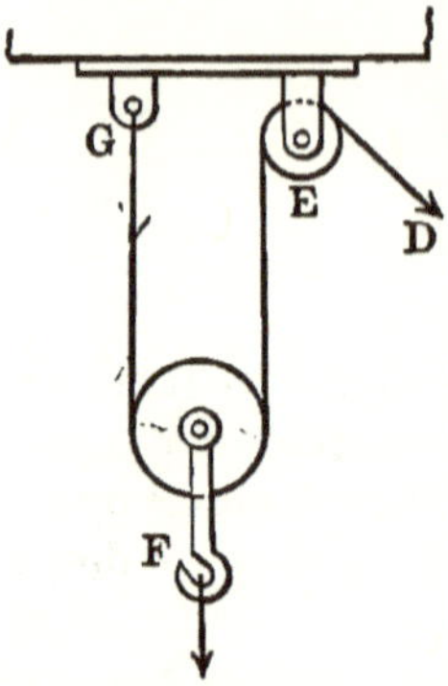

In Fig. 348 is a simple example where the rope is made fast at G, passes down around the pulley F, returns and passes over the pulley E, and on to some point D, where a pull may be exerted to raise a weight at F, as in the case of hay unloaders, stackers, etc.

FIG. 348.

When the ropes from F up are parallel, the velocity-ratio is constant; but when the rope diverges or converges upward, the velocity-ratio is variable, that is, the velocity of F as compared with that of D.

In this case D must move twice as far as F, and the pull exerted at F will be twice as great as at D.

Again, by making F the driver, D will be moved twice as far as F, and the pull exerted at D will be one half that at F, so that the

velocity-ratio $= 2$, or $1/2$.

In Fig. 349, F is raised faster than in Fig. 348. To find the velocity-ratio, draw *abcd* with lines parallel to the principal figure.

Then if D is pulled a distance cbd, F is raised a height ab, because ac is perpendicular to GF, and represents the movement of

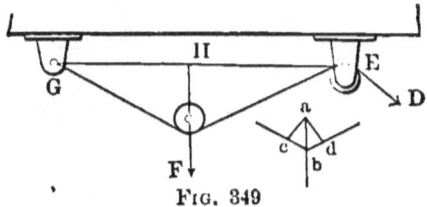

Fig. 349

a point of GF near F, when cbd is removed. Hence, for Fig. 349,

$$\frac{\text{velocity of } D}{\text{velocity of } F} = \frac{cbd}{ab} = \frac{2FH}{EF}.$$

Applying this to the case of Fig. 348, we will find $EF = FH$, and the velocity-ratio reduces to 2, as before stated. One make of a hay-unloader is as shown in Fig. 350, where the velocity-ratio is 3 to 1 and constant, when the ropes are parallel as shown.

In the Case of Ships' Tackle, where several sheaves are placed in a block with the ropes parallel, the velocity-ratio can readily be made out on the same principle as in the above figures. The ropes are usually nearly parallel in blocks and tackle, so that the velocity-ratio is nearly constant.

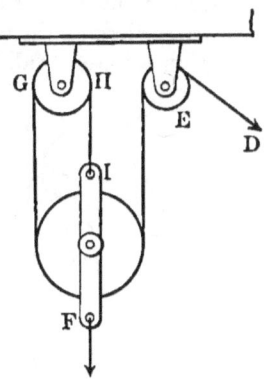

Fig. 350.

In Elevators operated by hydraulic power, the lifting rope passes from the cage up over a pulley at the top of the shaft or hoisting way, and down to a series of sheaves, and finally made fast to a stationary hitch at the lower end.

If, for instance, D, Fig. 351, represents the cage, G a set of stationary sheaves, F a set of sheaves attached to the hydraulic-power rod at F, then when F is drawn back D will be elevated.

Suppose the rope passes from the hitch near G to and around F, and then over to G, up to E, down to D, thus placing two ropes between the pulleys F and G. Then when F is drawn back 10 feet D will be elevated 20 feet, the multiplying being as many times as the ropes between G and F. The raising of D will be 80 feet for

8 ropes between *F* and *G*, and for a draft movement of 10 feet for *F*. The velocity-ratio will be 8 at the same time.

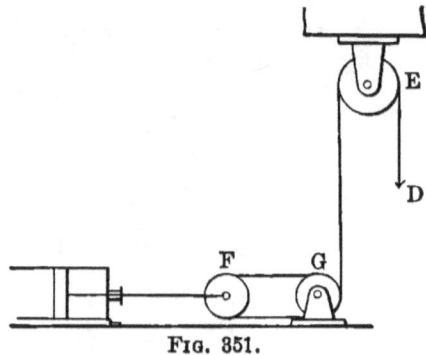

Fig. 351.

The Weston Differential Block, shown in Fig. 352, is a simple device for one so powerful.

An endless chain passes around *F*, then up and around *E*, then down in a loose "fall" *D*, then around *G*, slightly larger than *E*,

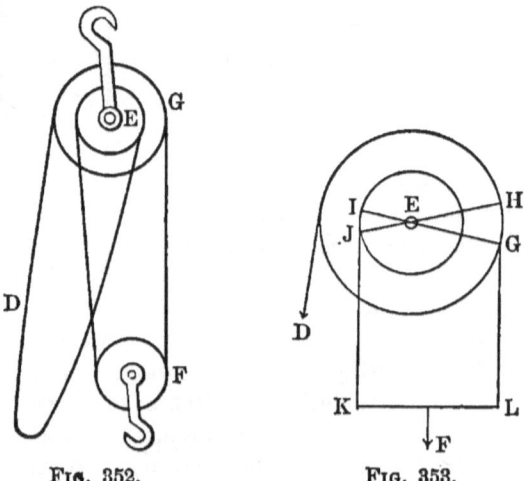

Fig. 352.   Fig. 353.

and down to *F* again; *G* and *E* being in one piece.

A chain is used instead of a rope, to serve as a hitch at *EG*, which are sprocketed to hold to the chain. Thus there is a hitch, in some measure equivalent to the rope hitch in the preceding cases, so that the device may be classed here.

Fig. 353 shows that when D is pulled a distance HG, L is raised by the amount GH, while K is lowered at the same time by the amount JI, and F is raised $1/2 \, (GH - IJ)$, so that the velocity-ratio, as between D and F, is

$$\frac{\text{velocity of } D}{\text{velocity of } F} = \frac{2HG}{HG - IJ} = \frac{2R}{R - r},$$

if R and r are the radii of the two sizes of sheaves, EG and EI.

If $R = 5''$ and $r = 4''$, then the velocity-ratio equals 10.

INDEX.

	PAGE
Addendum, dedendum, or root line............93, 107, 148, 161, 175, 178	
Alternate motions, circular, teeth for, limited........................	190
, unlimited........................	191
, limited and unlimited mangle-wheels..............	83–86
, pitch lines for.......................................	83
, teeth for...	123
Annular wheels, epicycloidal teeth for...............................	141
, interference.........................	141
, involute teeth.....................................	144
, interference.........................	144
Approximate gear teeth..	155–162

Belt gearing, any position of pulleys................................... 247
    , barrel and fusee for chronometers........................ 240
    , chain and sprocket 256
    , cone pulleys for lathes, etc 247
    , draw-bridge equalizer.................................... 239
    , for treadle.. 243
    , gas-meter prover equalizer............................... 238
    , law of motion given to find the wheels.................. 237
    , non-circular, for rifling machine......................... 241
    , pulley with high center................................. 246
    , quarter twist, guide pulley.............................. 246
    , retaining belt on pulley................................. 245
    , rope transmission....................................... 251–254
    , spinning-mule snail, etc................................. 241
    , the belt and circular pulley............................. 244
    , velocity-ratio, circular and non-circular.................. 236
Bevel gearing, circular, epicycloidal, involute 175
    , circular, pitch lines for.................................. 8
    , intermittent motion...................................... 82
    , non-circular pitch lines.................................. 62–74
    , teeth for non-circular wheels............................ 113

Cams, in general... 192
    , by co-ordinates...................................... 192
    intersection...................................... 193–195

INDEX.

	PAGE
Cams, conical	198
, roller for	216, 217
, cylindric, straight path	196
, curved "	197
, diameter constant	206, 213
, breadth "	207, 209
, easement for	203, 204
, headdle cam	202
, inverse, velocity-ratio	219
, law of motion defined	199–202
, return of follower by gravity	211
by spring, cam groove, positive	212
, roller, and its pin for	214, 217
, for spherical cam	217
, best form	216
, at salient points	215
, spherical	198
, tarrying points	202
, thick and irregular follower extremity	215
, uniform motion for	202
, velocity-ratio	194
, with flat-footed follower	199, 205, 208
several followers	208
Circular rolling wheels, pitch lines	7
, teeth for	128
Chain and sprocket gearing	256
Clearing curve and filleting	149
, possible	149
Conical cams	198
link-work	279
pulleys in belt gearing	247
roller for cams	217
wheels for gears, circular	8
Conjugate gear teeth for circular wheels	145–147
non-circular wheels	103
, Sang's theory for	168
Couplings, Almond	286
, axes not meeting, for	287
, Hooke's	283
, Oldham's	220–224
, Reuleaux's	288
Cutters for gear teeth, epicycloidal and involute	167
Cycle	3
Dead-points, in link-work	274
conic link-work	281
Directional relation	4
, constant, variable	4, 7

	PAGE
Elliptic wheels, pitch lines for...	30
———, interchangeable multilobes...	32–35
Epicycloidal curves, peculiar properties...	129
———— engine, to form gear-tooth curves...	164
———— gear teeth, annular wheels...	140
————————, flanks radial...	130
———————————— concave...	132
———————————— convex...	132, 140
————————, for inside pin gearing...	136
————————, interchangeable sets...	133, 140
————————, interference of annular gears...	141
————————, least crowding and friction...	137
————————, line of action...	151
————————, pin annular wheels...	136, 137
————————, and pin teeth...	134
————————, rack and pinion...	137–139
————————, two-tooth pinion...	136, 137
———— tooth cutters...	167
Escapements, anchor, for clocks, dead-beat...	226
————————, chronometer, for watches and chronometers...	234
————————, cylinder, for watches...	231
————————, duplex, " " ...	233
————————, gravity, for clocks, dead-beat...	229
————————, lever, for watches...	232
————————, pin-wheel, for clocks, dead-beat...	228
————————, recoil, " " ...	227
————————, power escapements...	225
Friction wheels, pitch lines for...	7
———— in cams...	213
————————, relieved by roller...	214
———— ratchets...	296
Grant's gear-cutting engine...	168, 170
Hindley's corset-shaped worm...	188
Hooke's universal joint or coupling...	283
Hyperbolic wheels, pitch lines...	36
Hyperboloids, circular rolling...	10
Intermittent motions, circular, locking arcs...	13
————————————, easements...	13
————————————, spurs and segments...	13
————————————, bevel and skew bevel...	189
————————————, non-circular, plane and bevel...	81, 82
————————————————, rolling spurs for...	120
————————————————, solid easement segments...	122–125
————————————————, teeth, spurs, and segments...	116–127
Involute bevel gear teeth...	175

	PAGE
Involute gear teeth, cutters for	167
, described with log-spiral	142
on drawing-board	155
, line of action for	151
Line of action of pressures between teeth	151
in belt gearing	237
centers, contact	5
Link-work, axes crossing without meeting	289
, or skew-bevel link-work, velocity-ratio	289
with sliding blocks	291, 292
of Willis	290
, axes parallel	259
, Corliss valve motion	260
, nailing machine motion	261
, needle bar motion	260
, conic, or solid, velocity-ratio	279
, Almond's	286
, examples	282–288
, gab and pin, and dead center	281
, for axes not meeting	287
, Hooke's joint, and velocity for	282, 283
, rolling wheel equivalent for	280
, Reuleaux's	288
, dead points in	274
, gabs and pins to pass	264, 275
, path for	276
, path and velocity of various points	261
, peculiar features	259
, rolling curve, equivalent for	264
examples	264–273
as ellipses	271
for crank and pitman	267
as hyperbolas	270
as link and drag link	268
for parabolas	272
Sylvester's kite	268
unsymmetrical, example	273
, sliding blocks and links for	263, 291
, velocity-ratio for	258
Machine, and parts of	1
-made gear teeth	164
gear-tooth cutters	166, 167
Mangle wheel and rack, pitch lines for	84–86
, teeth for	127
Mechanism, elementary combinations of	1
, primary and secondary trains of	1

INDEX.   307

	PAGE
Mechanism, train or trains of	1
, table of elementary combinations of	2
Motion, uniform, angular	3
Names and terms in mechanism	3
gearing	93
Non-circular wheels, plane	19
, equal log-spiral, segmental	20–24
, blocking tendency	107
, in general	44
, for intermittent motions	81
, internal or annular	111
, involute, teeth for	101
, laws of motion given, to find the wheels	48–53
, limit of eccentricity	106
, log-spiral wheels, complete	24–29
, one wheel given, to find its mate	44–47
, rolling, in extreme eccentricity	109
, five special forms	20
, solutions of practical problems	54–61
, bevel, drawn direct on normal sphere	73
, example	72
, in general, solution	69
, five special forms	62–66
, teeth for	112
, teeth for skew-bevel	114
Normal sphere, drawing wheels directly on	62–65, 71, 76
Odontograph, Grant's	161
, templet	157
, Willis'	158
Olivier spiraloid teeth for skew bevels	182
, contact between	183
, flat faces away from gorge	185
, interchangeability of	182
, interference of	183
, practical at gorge	187
, result of example	184
Parabolic wheels, equal and similar	35, 272
, equivalent link-work	272
, transformed	41
, with gabs and pins	276
Path of contact in gearing	149–150
, limited	151
Period	3
Pin and slit movement, inverse cam	219, 220
Pitch lines	7

INDEX.

	PAGE
Pitch lines, rolling in non-circular wheels	109, 110
, circumferential	94, 148
, diametral	94, 148
Point of contact	5, 7
, between the axes	7
, outside " "	8
Practical considerations in circular gearing	148–163
Rack and pinion, epicycloidal teeth	139
, involute "	144
Radius, rod to carry templet odontograph	157
tooth templet	154
Ratchet and click, movements	292
, form of teeth and click	295
, running	292
, continuous	295
, face ratchets	297
, friction ratchets	296
, wire feed	297
, reversible	294
, varied rate	294
, for varied steps	293
Return of follower in cams, by gravity	211
, by spring, by second cam, by groove	212
Reduplication, velocity-ratio	298
, elevator pulley and rope	300
, parallel ropes	298, 299
, ropes not parallel	298
, Weston's differential block	300
Revolution, a complete turn	4
Rolling contact	5
curve, equivalent for link work	264
hyperboloids	10, 75
Roller, in cam movements	214
, proper form of	216
, pin for	217
Rope transmission, short and long stretch	253
, for haulage lines	254
Rotation, a partial turn	4
Sang's theory of conjugating gear-teeth	168
Skew-bevel wheels, circular pitch surfaces	10–13
, error in early statement of	11
, non-circular to given law	76
, pin teeth	188
link-work	289
Sliding contact, in general, velocity-ratio	87–90
and rolling contact, in one model	90

INDEX.

	PAGE
Speeds, geometrical series of speeds in cone pulleys	247
Swasey's gear-cutting engine	168

Teeth of gear wheels, conjugate.................................. 102, 145
    , blocking tendency.................................. 107, 152
    , for bevel and skew-bevel non-circular.......... 112–115
        intermittent and alternate motions.......... 116–127
        involute non-circular.......................... 101
    , individually constructed.......................... 101
    , generation of tooth curves for.................. 92–105
        templets for...................... 95–101
    , limited inclination, hooking.................. 103–105
    , names and rules.................................. 93
    —, short, in eccentric wheels.................... 107, 152
    , trochoidal...................................... 99

Templets, for generating gear teeth, form and size.............. 95–100
    , motion templets.................................. 127
    , odontograph.................................... 157
    , path templets................................... 193
    , tooth   "   154

Tooth profile, epicycloidal, for circular wheels.................. 128, 152
    , involute,   "   "   " 143, 155
    , for non-circular wheels.......................... 98

Transformed wheels, and examples............................... 38
    , unilobed....................................... 39–42
    , bevel, several examples.......................... 66–68

Velocity, angular... 3
    , velocity-ratio................................... 3
        , in rolling contact............................ 5
    -ratio in belt-gearing............................ 237
        link-work................................ 258, 279, 289
        reduplication.................................. 298
        sliding contact................................ 87

www.ingramcontent.com/pod-product-compliance
Lightning Source LLC
Chambersburg PA
CBHW030012240426
43672CB00007B/924